A signpost stands at a fork in the road.
Pointing in one direction, the sign says "Victory."
Pointing in another direction, the sign says "Fulfillment."
We must pick a direction.
Which one will we choose?

If we choose the path to Victory,
the goal is to win!
We will experience the thrill of competition
as we rush toward the finish line.
Crowds gather to cheer for us!
And then it's over.
And everyone goes home.
(Hopefully we can do it again)

If we choose the path to Fulfillment,
The journey will be long.
There will be times in which we must watch our step
There will be times we can stop to enjoy the view
we keep going.
we keep going.
Crowds gather to join us on the journey.

And when our lives are over,
those who joined us on the path to Fulfillment
will keep going without us and
inspire others to join them too.

THE
INFINITE
GAME

BUSINESS

Simon Sinek is an optimist and the bestselling author of *Start With Why, Leaders Eat Last, Together Is Better* and *Find Your Why*. He is working to build a world in which the vast majority of us will wake up inspired, feel safe at work and return home fulfilled at the end of the day. His TED talk, 'How Great Leaders Inspire Action' is one of the most widely viewed of all time.

THE
INFINITE
GAME

THE
INFINITE
GAME
SIMON
SINEK

THE
INFINITE
GAME
THE
INFINITE
GAME
THE

BUSINESS

PENGUIN BUSINESS

UK | USA | Canada | Ireland | Australia
India | New Zealand | South Africa

Penguin Business is part of the Penguin Random House group of companies
whose addresses can be found at global.penguinrandomhouse.com.

First published in the United States of America by Portfolio/Penguin,
an imprint of Penguin Random House LLC 2019
First published in Great Britain by Penguin Business 2019
This edition published 2020
001

Copyright © SinekPartners LLC, 2019

Printed and bound in Great Britain by Clays Ltd, Elcograf S.p.A.

A CIP catalogue record for this book is available from the British Library

ISBN: 978-0-241-38563-0

Follow us on LinkedIn: https://www.linkedin.com/company/penguin-connect/

www.greenpenguin.co.uk

Dear Grandma,

Because you lived as if there was no finish line.
May we all learn to live such an infinite life.

Love, Simon

CONTENTS

WHY I WROTE THIS

I t's surprising that this book even needs to exist. Over the course of human history, we have seen the benefits of infinite thinking so many times. The rise of great societies, advancements in science and medicine and the exploration of space all happened because large groups of people, united in common cause, chose to collaborate with no clear end in sight. If a rocket that was headed for the stars crashed, for example, we figured out what was wrong and tried again . . . and again . . . and again. And even after we succeeded, we kept going. We did these things not because of the promise of an end-of-year bonus; we did these things because we felt like we were contributing to something bigger than ourselves, something with value that would last well beyond our own lifetimes.

For all its benefits, acting with an infinite, long-term view is not easy. It takes real effort. As human beings we are naturally inclined to seek out immediate solutions to uncomfortable problems and prioritize quick wins to advance our ambitions. We tend to see the world in terms of successes and failures, winners and losers. This default win-lose mode can sometimes work for the short term; however, as a strategy for how companies and organizations operate, it can have grave consequences over the longer term.

The results of this default mindset are all too familiar: annual rounds of mass layoffs to meet arbitrary projections, cutthroat work environments, subservience to the shareholder over the needs of employees and customers, dishonest and unethical business practices, rewarding high-performing toxic team members while turning a blind eye to the damage they are doing to the rest of the team and rewarding leaders who seem to care a lot more about themselves than those in their charge. All things that contribute to a decline of loyalty and engagement and an increase of insecurity and anxiety that too many of us feel these days. This impersonal and transactional approach to business seems to have accelerated in the aftermath of the Industrial Revolution and seems to be accelerating even more in our digital age. Indeed, our entire understanding of commerce and capitalism seems to have fallen under the sway of short-term, finite-minded thinking.

Though many of us lament this state of things, unfortunately it seems like the market's desire to maintain the status quo is more powerful than the momentum to change it. When we say things like "people must come before profit," we often face resistance. Many of those who control the current system, many of our current leaders, tell us we are naïve and don't understand the "reality" of how business works. As a result, too many of us back down. We resign ourselves to waking up dreading to go to work, not feeling safe when we are there and struggling to find fulfillment in our lives. So much so that the search for that elusive work-life balance has become an entire industry unto itself. It leaves me wondering, do we have another, viable option?

It is entirely possible that perhaps, just perhaps, the "reality" the cynics keep talking about doesn't have to be that way. That perhaps our current system of doing business isn't "right," or even "best." It is just the system that we are used to, one preferred and advanced by a minority, not the majority. If this is, indeed, the case, then we have an opportunity to advance a different reality.

It is well within our power to build a world in which the vast majority of us wake up every single morning inspired, feel safe at work and return home fulfilled at the end of the day. The kind of change I advocate is not easy. But it is possible. With good leaders—great leaders—this vision can come to life. Great leaders are the ones who think beyond "short term" versus "long term." They are the ones who know that it is not about the next quarter or the next election; it is about the next generation. Great leaders set up their organizations to succeed beyond their own lifetimes, and when they do, the benefits—for us, for business and even for the shareholder— are extraordinary.

I wrote this book not to convert those who defend the status quo, I wrote this book to rally those who are ready to challenge that status quo and replace it with a reality that is vastly more conducive to our deep-seated human need to feel safe, to contribute to something bigger than ourselves and to provide for ourselves and our families. A reality that works for our best interests as individuals, as companies, as communities and as a species.

If we believe in a world in which we can feel inspired, safe and fulfilled every single day and if we believe that leaders are the ones who can deliver on that vision, then it is our collective responsibility to find, guide and support those who are committed to leading in a way that will more likely bring that vision to life. And one of the steps we need to take is to learn what it means to lead in the Infinite Game.

Simon Sinek
February 4, 2019
London, England

THE
INFINITE
GAME

WINNING

On the morning of January 30, 1968, North Vietnam launched a surprise attack against U.S. and allied forces. Over the next twenty-four hours, more than 85,000 North Vietnamese and Viet Cong troops attacked over 125 targets across the country. The American forces were caught completely off guard. So much so that many of the commanding officers weren't even at their posts when the attacks began—they were away celebrating Tết in nearby cities. The Tết Offensive had begun.

Tết is the Lunar New Year and it is as significant to the Vietnamese as Christmas is to many Westerners. And, like the Christmas truce of World War I, there was a decades-old tradition in Vietnam that there was never any fighting on Tết. However, seeing an opportunity to overwhelm American forces and hopefully bring a swift end to the war, North Vietnamese leadership decided to break with tradition when they launched their surprise offensive.

Here's the amazing thing: the United States repelled every single attack. Every single one. And American troops didn't just repel the onslaughts, they decimated the attacking forces. After most of the major fighting had come to an end, about a week after the initial attack, America had lost fewer than a thousand troops. North Vietnam, in stark contrast, lost over 35,000 troops! In the city of Hué, where fighting continued for almost a month, America lost 150 Marines compared to an estimated 5,000 troops the North Vietnamese lost!

A close examination of the Vietnam War as a whole reveals a remarkable picture. America actually won the vast majority of the battles it fought. Over the course of the ten years in which U.S. troops were active in the Vietnam War, America lost 58,000 troops. North Vietnam lost over 3 million people. As a percent of population, that's the equivalent of America losing 27 million people in 1968.

All this begs the question, how do you win almost every battle, decimate your enemy and still lose the war?

Chapter 1

FINITE AND INFINITE GAMES

I f there are at least two players, a game exists. And there are
two kinds of games: finite games and infinite games.

Finite games are played by known players. They have
fixed rules. And there is an agreed-upon objective that, when
reached, ends the game. Football, for example, is a finite
game. The players all wear uniforms and are easily identifiable. There is a set of rules, and referees are there to enforce
those rules. All the players have agreed to play by those rules
and they accept penalties when they break the rules. Everyone agrees that whichever team has scored more points by
the end of the set time period will be declared the winner,
the game will end and everyone will go home. In finite
games, there is always a beginning, a middle and an end.

Infinite games, in contrast, are played by known and

unknown players. There are no exact or agreed-upon rules. Though there may be conventions or laws that govern how the players conduct themselves, within those broad boundaries, the players can operate however they want. And if they choose to break with convention, they can. The manner in which each player chooses to play is entirely up to them. And they can change how they play the game at any time, for any reason.

Infinite games have infinite time horizons. And because there is no finish line, no practical end to the game, there is no such thing as "winning" an infinite game. In an infinite game, the primary objective is to keep playing, to perpetuate the game.

My understanding of these two types of games comes from the master himself, Professor James P. Carse, who penned a little treatise called *Finite and Infinite Games: A Vision of Life as Play and Possibility* in 1986. It was Carse's book that first got me thinking beyond winning and losing, beyond ties and stalemates. The more I looked at our world through Carse's lens of finite and infinite games, the more I started to see infinite games all around us, games with no finish lines and no winners. There is no such thing as coming in first in marriage or friendship, for example. Though school may be finite, there is no such thing as winning education. We can beat out other candidates for a job or promotion, but no one is ever crowned the winner of careers. Though nations may compete on a global scale with other nations for land, influence or economic advantage, there is no such thing as winning global politics. No matter how successful we are in life, when we die, none of us will be declared the winner of life. And there is certainly no such thing as winning business. All these things are journeys, not events.

However, if we listen to the language of so many of our leaders today, it's as if they don't know the game in which they are playing. They talk constantly about "winning." They obsess about "beating their competition." They announce to the world that they are "the best." They state that their vision is to "be number one." Except that in games without finish lines, all of these things are impossible.

When we lead with a finite mindset in an infinite game, it leads to all kinds of problems, the most common of which include the decline of trust, cooperation and innovation. Leading with an infinite mindset in an infinite game, in contrast, really does move us in a better direction. Groups that adopt an infinite mindset enjoy vastly higher levels of trust, cooperation and innovation and all the subsequent benefits. If we are all, at various times, players in infinite games, then it is in our interest to learn how to recognize the game we are in and what it takes to lead with an infinite mindset. It is equally important for us to learn to recognize the clues when finite thinking exists so that we can make adjustments before real damage is done.

The Infinite Game of Business

The game of business fits the very definition of an infinite game. We may not know all of the other players and new ones can join the game at any time. All the players determine their own strategies and tactics and there is no set of fixed rules to which everyone has agreed, other than the law (and even that can vary from country to country). Unlike a finite game, there is no predetermined beginning, middle or end to business. Although many of us agree to certain time frames for evaluating our own performance relative to that of other players—the financial year, for example—those time frames represent markers within the course of the game; none marks the end of the game itself. The game of business has no finish line.

Despite the fact that companies are playing in a game that cannot be won, too many business leaders keep playing as if they can. They continue to make claims that they are the "best" or that they are "number one." Such claims have become so commonplace that we rarely, if ever, stop to actually think about how ridiculous some of them are. Whenever I see a company claim that it is number one or the best, I always like to look at the fine print to see how they cherry-picked the metrics. For years, British Airways, for example, claimed in their advertising that they were "the world's favourite airline." Richard Branson's airline, Virgin Atlantic, filed a dispute with Britain's Advertising Standards Authority that such a claim could not be true based on recent passenger surveys. The ASA allowed the claim to stand, however, on the basis that British Airways carried more international passengers than any other airline. "Favourite," as they used the word, meant that their operation was expansive, not necessarily preferred.

To one company, being number one may be based on the number of customers they serve. To another, it could be about revenues, stock performance, the number of employees or the number of offices they have around the globe. The companies making the claims even get to decide the time frames in which they are making their calculations. Sometimes it's a quarter. Or eight months. Sometimes a year. Or five years. Or a dozen. But did everyone else in their industry agree to those same time frames for comparison? In finite games, there's a single, agreed-upon metric that separates the winner from the loser, things like goals scored, speed or strength. In infinite games, there are multiple metrics, which is why we can never declare a winner.

In a finite game, the game ends when its time is up and the players live on to play another day (unless it was a duel, of course). In an infinite game, it's the opposite. It is the game that lives on and it is the players whose time runs out. Because there is no such thing as winning or losing in an infinite game, the players simply drop out of the game when they run out of the will and resources to keep playing. In business we call this bankruptcy or sometimes merger or acquisition. Which means, to succeed in the Infinite Game of business, we have to stop thinking about who wins or who's the best and start thinking about how to build organizations that are strong enough and healthy enough to stay in the game for many generations to come. The benefits of which, ironically, often make companies stronger in the near term also.

A Tale of Two Players

Some years ago, I spoke at an education summit for Microsoft. A few months later, I spoke at an education summit for Apple. At the Microsoft event, the majority of the presenters devoted a good portion of their presentations to talking about how they were going to beat Apple. At the Apple event, 100 percent of the presenters spent 100 percent of their time talking about how Apple was trying to help teachers teach and help students learn. One group seemed obsessed with beating their competition. The other group seemed obsessed with advancing a cause.

After my talk at Microsoft, they gave me a gift—the new Zune (when it was still a thing). This was Microsoft's answer to Apple's iPod, the dominant player in the MP3-player market at the time. Not to be outdone, Microsoft introduced the Zune to try to steal market share from their archrival. Though he knew it wouldn't be easy, in 2006, then CEO of Microsoft Steve Ballmer was confident that Microsoft could eventually "beat" Apple. And if the quality of the product was the only factor, Ballmer was right to be optimistic. The version Microsoft gave me—the Zune HD—was, I have to admit, quite exceptional. It was elegantly designed. The user interface was simple, intuitive and user-friendly. I really, really liked it. (In the interest of full disclosure, I gave it away to a friend for the simple reason that unlike my iPod, which was compatible with Microsoft Windows, the Zune was not compatible with iTunes. So as much as I wanted to use it, I couldn't.)

After my talk at the Apple event, I shared a taxi back to the hotel with a senior Apple executive, employee number 54 to be exact, meaning he'd been at the company since the early days and was completely immersed in Apple's culture and belief set. Sitting there with him, a captive audience, I couldn't help myself. I had to stir the pot a little. So I turned to him and said, "You know . . . I spoke at Microsoft and they gave me their new Zune, and I have to tell you, it is *SO MUCH BETTER* than your iPod touch." The executive looked at me, smiled, and replied, "I have no doubt." And that was it. The conversation was over.

The Apple exec was unfazed by the fact that Microsoft had a better product. Perhaps he was just displaying the arrogance of a dominant market leader. Perhaps he was putting on an act (a very good one). Or perhaps there was something else at play. Although I didn't know it at the time, his response was consistent with that of a leader with an infinite mindset.

The Benefits of an Infinite Mindset

In the Infinite Game, the true value of an organization cannot be measured by the success it has achieved based on a set of arbitrary metrics over arbitrary time frames. The true value of an organization is measured by the desire others have to contribute to that organization's ability to keep succeeding, not just during the time they are there, but well beyond their own tenure. While a finite-minded leader works to get something from their employees, customers and shareholders in order to meet arbitrary metrics, the infinite-minded leader works to ensure that their employees, customers and shareholders remain inspired to continue contributing with their effort, their wallets and their investments. Players with an infinite mindset want to leave their organizations in better shape than they found them. Lego invented a toy that has stood the test of time not because it was lucky, but because nearly everyone who works there wants to do things to ensure that the company will survive them. Their drive is not to beat the quarter, their drive is to "continue to create innovative play experiences and reach more children every year."

According to Carse, a finite-minded leader plays to end the game—to win. And if they want to be the winner, then there has to be a loser. They play for themselves and want to defeat the other players. They make every plan and every move with winning in mind. They almost always believe they *must* act that way, even though, in fact, they don't have to at all. There is no rule that says they have to act that way. It is their mindset that directs them.

Carse's infinite player plays to keep playing. In business, that means building an organization that can survive its leaders. Carse also expects the infinite player to play for the good of the game. In business, that means seeing beyond the bottom line. Where a finite-minded player makes products they think they can sell to people, the infinite-minded player makes products that people want to buy. The former is primarily focused on how the sale of those products benefits the company; the latter is primarily focused on how the products benefit those who buy them.

Finite-minded players tend to follow standards that help them achieve their personal goals with less regard to the effects of the ripples that may cause. To ask, "What's best for me" is finite thinking. To ask, "What's best for us" is infinite thinking. A company built for the Infinite Game doesn't think of itself alone. It considers the impact of its decisions on its people, its community, the economy, the country and the world. It does these things for the good of the game. George Eastman, the founder of Kodak, was devoted to his vision of making photography easy and accessible to everyone. He also recognized that advancing his vision was intimately tied to the well-being of his people and the community in which they lived. In 1912, Kodak was the first company to pay employees a dividend based on company performance and several years later issued what we now know as stock options. They also provided their employees with a generous benefits package, gave paid time off for sick leave (it was a new idea then) and subsidized tuitions for employees who took classes at local colleges. (All things that have been adopted by many other companies. In other words, it was not only good for Kodak, it was good for the game of business.) In addition to the tens of thousands of jobs Kodak provided, Eastman built a hospital, founded a music school, and gave generously to institutions of higher learning, including the Mechanics Institute of Rochester (which was later renamed

Rochester Institute of Technology) and the University of Rochester.

Because they are playing with an end point in mind, Carse tells us, finite-minded players do not like surprises and fear any kind of disruption. Things they cannot predict or cannot control could upset their plans and increase their chances of losing. The infinite-minded player, in contrast, expects surprises, even revels in them, and is prepared to be transformed by them. They embrace the freedom of play and are open to any possibility that keeps them in the game. Instead of looking for ways to react to what has already happened, they look for ways to do something new. An infinite perspective frees us from fixating on what other companies are doing, which allows us to focus on a larger vision. Instead of reacting to how new technology will challenge our business model, for example, those with infinite mindsets are better able to foresee the applications of new technology.

It's easy now to see why the Apple executive with whom I shared a cab could be so nonchalant about Microsoft's well-designed Zune. He understood that, in the Infinite Game of business, sometimes Apple would have the better product, sometimes another company would have the better product. They weren't trying to outdo Microsoft; Apple was trying to outdo itself. The company was looking ahead to what would come after the iPod. Apple's infinite mindset helped them think, not outside the box, but beyond it. About a year after the Zune was first introduced, Apple released the first iPhone. The iPhone redefined the entire category of smartphones and rendered both the Zune and the iPod virtually obsolete. Though some people believed Apple could predict consumer preferences and see into the future, they couldn't. In reality it was their infinite perspective that opened a path for them to innovate in ways that companies with more finite-minded leadership simply could not.

A finite-focused company may come up with "innovative" ways to boost the bottom line, but those decisions don't usually benefit the organization, the employees, the customers and the community—those who exist beyond the bottom line. Nor do they necessarily leave the organization in better shape for the future. And the reason is simple. It's because those decisions tend to be made primarily for the benefit of the people who made them and not with the infinite future in mind . . . just the near future. In contrast, infinite-minded leaders don't ask their people to fixate on finite goals; they ask their people to help them figure out a way to advance toward a more infinite vision of the future that benefits everyone. The finite goals become the markers of progress toward that vision. And when everyone focuses on the infinite vision, it not only drives innovation, but it also drives up the numbers. Indeed, companies led by infinite-minded leaders often enjoy record-making profits. What's more, the inspiration, innovation, cooperation, brand loyalty and profits that result from infinite-minded leadership serve companies not just in times of stability but also in times of instability. The same things that help the company survive and thrive during good times help make the company strong and resilient in hard times.

A company built for resilience is a company that is structured to last forever. This is different from a company built for stability. Stability, by its very definition, is about remaining the same. A stable organization can theoretically weather a storm, then come out of it the same as it was before. In more practical terms, when a company is described as stable, it is usually to draw a contrast to another company that is higher risk and higher performing. "Slow growth but stable," so goes the thinking. But a company built for stability still fails to understand the nature of the Infinite Game, for it is likely still not prepared for the unpredictable—for the new technology, new competitor, market shift or world events that can, in an instant, derail their strategy. An infinite-minded leader does not simply want to build a company that can weather change but one that can be transformed by it. They want to build a company that embraces surprises and adapts with them. Resilient companies may come out the other end of upheaval entirely different than they were when they went in (and are often grateful for the transformation).

Victorinox, the Swiss company that made the Swiss Army knife famous, saw its business dramatically affected by the events of September 11. The ubiquitous corporate promotional item and standard gift for retirements, birthdays and graduations, in an instant, was banned from our hand luggage. Whereas most companies would take a defensive posture—fixating on the blow to their traditional model and how much it was going to cost them—Victorinox took the offense. They embraced the surprise as an opportunity rather than a threat—a characteristic move of an infinite-minded player. Rather than employing extreme cost cutting and laying off their workforce, the leaders of Victorinox came up with innovative ways to save jobs (they made no layoffs at all), increased investment in new product development and inspired their people to imagine how they could leverage their brand into new markets.

In good times, Victorinox built up reserves of cash, knowing that at some point there would be more difficult times. As CEO Carl Elsener says, "When you look at the history of world economics, it was always like this. Always! And in the future, it will always be like this. It will never go only up. It will never go only down. It will go up and down and up and down. . . . We do not think in quarters," he says. "We think in generations." This kind of infinite thinking put Victorinox in a position where they were both philosophically and financially ready to face what for another company might have been a fatal crisis. And the result was astonishing. Victorinox is now a different and even stronger company than it was before September 11. Knives used to account for 95 percent of the company's total sales (Swiss Army knives alone accounted for 80 percent). Today, Swiss Army knives account for only 35 percent of total revenue, but sales of travel gear, watches and fragrances have helped Victorinox nearly double its revenues compared to the days before September 11. Victorinox is not a stable company, it is a resilient one.

The benefits of playing with an infinite mindset are clear and multifaceted. So what happens when we play with a finite mindset in the Infinite Game of business?

The Detriments of a Finite Mindset in an Infinite Game

Decades after the Vietnam War, Robert McNamara, U.S. Secretary of Defense during the war, had the chance to meet Nguyen Co Thach, the North Vietnamese Foreign Ministry's chief specialist on the United States from 1960 to 1975. McNamara was flabbergasted by how badly America misunderstood their enemy. "You must never have read a history book," McNamara recounts Thach scolding him. "If you had, you'd know we weren't pawns of the Chinese or the Russians. . . . Don't you understand that we have been fighting the Chinese for a thousand years?" Thach went on. "We were fighting for our independence! And we would fight to the last man! And we were determined to do so! And no amount of bombing, no amount of U.S. pressure would ever have stopped us!" The North Vietnamese were playing an infinite game with an infinite mindset.

The United States assumed the Vietnam War was finite because most wars are, indeed, finite. In most wars there is a land grab or some other easy to measure finite objective. If the combatants enter the war with clear political objectives, whoever achieves their finite objective first will be declared victor, a treaty will be signed and the war will end. But that's not always the case. Had America's leaders paid closer attention, perhaps they would have recognized the true nature of the Vietnam War sooner. There were clues all around.

For starters, there was no clear beginning, middle and end to America's involvement in Vietnam. Nor was there a clear political objective that, when achieved, would allow them to declare victory and bring their troops home. And even if there had been, the North Vietnamese would not have agreed to it. The Americans also seem to have misunderstood who they were fighting against. They believed the conflict in Vietnam was a proxy war against China and the Soviet Union. But the North Vietnamese were ardent that they were no puppet of any other government. Vietnam had been fighting against imperialist influence for decades, against the Japanese during World War II, then against the French afterward. To the North Vietnamese, the war with the United States wasn't an extension of the Cold War; it was a fight against yet another interventionist power. Even the manner in which the North Vietnamese fought—their propensity to disobey the conventions of traditional warfare and their will to keep fighting no matter how many people they lost—should have signaled to America's leaders that they had misjudged the nature of the game they were in.

When we play with a finite mindset in an infinite game, the odds increase that we will find ourselves in a quagmire, racing through the will and resources we need to keep playing. And this is what happened to America in Vietnam. The United States operated as if the game were finite instead of fighting against a player that was playing with the right mindset for the Infinite Game they were actually in. While America was fighting to "win," the North Vietnamese were fighting for their lives! And both made strategic choices according to their mindset. Despite their vastly superior military might, there was simply no way the United States could prevail. What brought America's involvement in Vietnam to an end was not a military or political win or loss, but public pressure back home. The American people could no longer support a seemingly unwinnable and expensive war in a faraway land. It's not that America "lost" the Vietnam War, rather it had exhausted the will and resources to keep playing . . . and so it was forced to drop out of the game.

The Quagmire of Vietnam in Business

When Microsoft launched the Zune, there was no grand vision that the product was helping to advance. They weren't thinking about what possibilities the future might hold. It was just a competition for market share and money—one in which Microsoft wasn't doing very well. Ballmer's prediction that the Zune could "beat" the iPod couldn't have been more wrong. Debuting with a 9 percent market share, the Zune's popularity declined steadily until it hit 1 percent in 2010. The following year it was discontinued. The iPod, in contrast, enjoyed around a 70 percent slice for the same time period.

Some have argued that the Zune failed because Microsoft didn't invest enough in advertising. But the theory doesn't hold up. Spanx, Sriracha, and GoPro are just three brands that relied solely on word of mouth to increase brand awareness. All three not only emerged from obscurity without traditional advertising, but went on to thrive without it. Others suggest that the Zune failed because Microsoft was too late to the MP3 player market. This theory doesn't hold up much better. Apple itself introduced the iPod a full five years after MP3 players were a well-known product category. Brands like Rio, Nomad and Sony were already advancing the technology and selling well. Yet, within four years of its 2001 launch, the iPod had gained the lion's share of the U.S. digital music player market . . . a number that only continued to rise.

As great as Microsoft's Zune may have been, it wasn't the design, marketing or the timing of the product that were the problem. It takes more than all those things to survive and thrive in the Infinite Game of business. Great products fail all the time. How a company is led must also be considered. Prioritizing comparison and winning above all else, finite-minded leaders will set corporate strategy, product strategy, incentive structures and hiring decisions to help meet finite goals. And with a finite mindset firmly entrenched in almost all aspects of the organization, a sort of tunnel vision results. The result of which pushes almost everyone inside the company to place excessive focus on the urgent at the expense of the important. Executives instinctively start to respond to known factors instead of exploring or advancing unknown possibilities. And in some cases, leaders can become so obsessed with what the competition is doing, falsely believing they need to react to their every move, that they become blind to a whole host of better choices to strengthen their own organization. It's like trying to win by playing defense. Seduced by a finite mindset, Microsoft found themselves in a never-ending game of whack-a-mole.

Microsoft's leaders failed to appreciate the Infinite Game they were in and the infinite mindset with which Apple was playing. Though Steve Ballmer sometimes spoke of "vision" or the "long term," like other finite-minded leaders who use this kind of infinite language, he almost always did so in the finite context of rank, stock performance, market share and money. Playing with the wrong mindset for the game they were in, Microsoft was chasing an impossible objective—"to win." Wasting the will and resources needed to stay in the game, like America in Vietnam, Microsoft was in quagmire.

It seemed the company had not learned its lesson with the iPod. When the iPhone came out in 2007, Ballmer's reaction to it underscored his finite perspective. Questioned about the iPhone in an interview, he scoffed, "There's no chance that the iPhone is going to get any significant market share. No chance. . . . They may make a lot of money. But if you actually take a look at the 1.3 billion phones that get sold, I'd prefer to have our software in 60% or 70% or 80% of them, than I would to have 2% or 3%, which is what Apple might get." Constrained by a finite mindset, Ballmer was more focused on the relative numbers the iPhone could achieve instead of seeing how it might alter the entire market . . . or even completely redefine the role our phones play in our lives. In a turn of events that must have driven Ballmer crazy, after just five years on the market, iPhone sales alone were higher than all of Microsoft's products combined.

In 2013, at his final press conference as CEO of Microsoft, Steve Ballmer summed up his career in a most finite-minded way. He defined success based on the metrics he selected within the time frame of his own tenure in the job. "In the last five years, probably Apple has made more money than we have," he said. "But in the last thirteen years, I bet we've made more money than almost anybody on the planet. And that, frankly, is a great source of pride to me." It seems Ballmer was trying to say that under the thirteen years of his leadership, his company had "won." Imagine how different that press conference could have been if, instead of looking back at a balance sheet, Ballmer shared all the things Microsoft had done and could still do to advance Bill Gates's original infinite vision: "To empower every person and every organization on the planet to achieve more."

A finite-minded leader uses the company's performance to demonstrate the value of their own career. An infinite-minded leader uses their career to enhance the long-term value of the company . . . and only part of that value is counted in money. The game didn't end simply because Ballmer retired. The company continued to play without him. In the Infinite Game, how well he did financially is much less important than whether he left the company culture adequately prepared to survive and thrive for the next thirteen years. Or thirty-three years. Or three hundred years. And on that standard, Ballmer lost.

In the Infinite Game of business, when our leaders maintain a finite mindset or put too much focus on finite objectives, they may be able to achieve a number one ranking with an arbitrary metric over an arbitrary time frame. But that doesn't necessarily mean they are doing the things they need to ensure that the company can keep playing for as long as possible. In fact, more often than not, the things they do harm the company's inner workings and, without intervention, accelerate the company's ultimate demise.

Because finite-minded leaders place unbalanced focus on near-term results, they often employ any strategy or tactic that will help them make the numbers. Some favorite options include reducing investment in research and development, extreme cost cutting (e.g., regular rounds of layoffs, opting for cheaper, lower quality ingredients in products, cutting corners in manufacturing or quality control), growth through acquisition and stock buybacks. These decisions can, in turn, shake a company's culture. People start to realize that nothing and no one is safe. In response, some instinctually behave as if they were switched to self-preservation mode. They may hoard information, hide mistakes and operate in a more cautious, risk-averse way. To protect themselves, they trust no one. Others double down on an only-the-fittest-survive mentality. Their tactics can become overly aggressive. Their egos become unchecked. They learn to manage up the hierarchy to garner favor with senior leadership while, in some cases, sabotaging their own colleagues. To protect themselves, they trust no one. Regardless of whether they are in self-preservation or self-promotion mode, the sum of all of these behaviors contributes to a general decline in cooperation across the company, which also leads to stagnation of any truly new or innovative ideas. This is what happened at Microsoft.

Consumed by the finite game, Microsoft became obsessed with quarterly numbers. Many of the people who had been at the company from the early days lamented a loss of inspiration, imagination and innovation. Trust and cooperation suffered as internal product groups started to fight with each other instead of supporting each other. And as if large companies don't struggle enough with silos, Microsoft's divisions sometimes actively worked to undermine each other. It went from being a place that attracted people to join a crusade to a place that the best and brightest avoided like the plague. A company that used to be a "lean competition machine led by young visionaries of unparalleled talent," as *Vanity Fair* reported, "mutated into something bloated and bureaucracy-laden, with an internal culture that unintentionally rewards managers who strangle innovative ideas that might threaten the established order of things." In other words, a finite mindset left the company culture a mess.

It can take a long time for very large companies with a finite-minded leader at the helm to exhaust the will and resources accumulated by the infinite leader that preceded them. Under Ballmer, Microsoft was still a dominant player, especially in business markets. This was largely thanks to the groundwork laid under the more infinite-minded Bill Gates. Had Ballmer stayed, or another finite leader replaced him, however, the will of the people to keep fighting the good fight and the resources the company would need to keep playing would eventually have run out. Just because a company is big and has enjoyed financial success does not mean it is strong enough to last.

Microsoft's experience is not unique. Business history is littered with similar cautionary tales. General Motors' obsession with market share over profit, for example, would have put them out of business if it weren't for a government bailout. Sears, Circuit City, Lehman Brothers, Eastern Airlines and Blockbuster Video were not so lucky. They are just a few more examples of once strong, well-established companies whose leaders were seduced by the thrill of playing with a finite mindset only to put their companies on a path to destruction.

Sadly, over the course of the past thirty to forty years, finite-minded leadership has become the modern standard in business. Finite-minded leadership is embraced by Wall Street and taught in business schools. At the same time, the life span of companies appears to be getting shorter and shorter. According to a study by McKinsey, the average life span of an S&P 500 company has dropped over forty years since the 1950s, from an average of sixty-one years to less than eighteen years today. And according to Professor Richard Foster of Yale University, the rate of change "is at a faster pace than ever." I accept there are multiple factors that contribute to these numbers, but we must consider that too many leaders today are building companies that are simply not made to last. Which is ironic because even the most goal-oriented, finite-minded leader must concede that the longer an organization can survive and thrive, the more likely it is to achieve *all* its goals.

It's not just companies that are impacted by too much finite-minded leadership. With more finite thinkers in positions of authority in all facets of life comes increased pressure to change public policy to further entrench even more finite-mindedness. And before too long, we have an entire economy operating within the constraints of a finite mindset, playing by the rules for a game we are not in. This is an untenable situation. And the data reflects it. After the 1929 stock market crash that lead to the Great Depression, for example, the Glass-Steagall Act was introduced to curb some of the more finite-minded corporate behaviors that were the cause of the instability in the markets at that time. Between the time Glass-Steagall was passed until the 1980s and '90s, when the act was virtually gutted in the name of opening up the financial markets, the number of stock market crashes that happened was zero. Since the gutting, however, we have had three: Black Monday in 1987, the burst of the dot-com bubble in 2000 and the financial crisis of 2008.

When we play with a finite mindset in the Infinite Game, we will continue to make decisions that sabotage our own ambitions. It's like eating too many desserts in the name of "enjoying life" only to make oneself diabetic in the process. Creating the conditions for a stock market crash are an extreme example of what happens when too many players in the game opt to play with a finite mindset. The more likely scenario is a general decline in trust, cooperation and innovation in an organization, all of which make it vastly more difficult to survive and thrive in a fast-moving business world. If we believe trust, cooperation and innovation matter to the long-term prospects of our organizations, then we have only one choice—to learn how to play with an infinite mindset.

Lead with an Infinite Mindset

There are three factors we must always consider when deciding how we want to lead:

1. We don't get to choose whether a particular game is finite or infinite.
2. We do get to choose whether or not we want join the game.

3. Should we choose to join the game, we can choose whether we want to play with a finite or an infinite mindset.

If we join a finite game, clearly we want to play by the right rules in order to increase our chances of winning. There is no use preparing to play basketball if we are about to enter a game of football. The same is true if we decide to become a leader in an infinite game. We are more likely to survive and thrive if we play for the game we are in.

The choice to lead with an infinite mindset is less like preparing for a football game and more like the decision to get into shape. There is no one thing we can do in order to get into shape. We can't simply go to the gym for nine hours and expect to be in shape. However, if we go to the gym every single day for twenty minutes, we will absolutely get into shape. Consistency becomes more important than intensity. The problem is, no one knows exactly when we will see results. In fact, different people will show results at different times. But without question, 100 percent, we all know it will work. And though we may have finite fitness goals we want to reach, if we want to be as healthy as possible, the lifestyle we adopt matters more than whether or not we hit our goal on the arbitrary dates we set. With any health regime, there are certain things we have to do—eat more vegetables, work out on a regular basis and get enough sleep, for example. Adopting an infinite mindset is exactly the same.

Any leader who wants to adopt an infinite mindset must follow five essential practices:

- Advance a Just Cause
- Build Trusting Teams

- Study your Worthy Rivals
- Prepare for Existential Flexibility
- Demonstrate the Courage to Lead

If we want to follow a health regime, we can choose to follow some of the practices but not all of them—we can exercise but never eat vegetables, for example. If we choose this approach, we may get some benefit. But we will only enjoy the full benefit if we do everything. Likewise, there is a benefit to following some of the practices required for infinite thinking. However, to fully equip an organization for a long and healthy life in the Infinite Game, we must do it all.

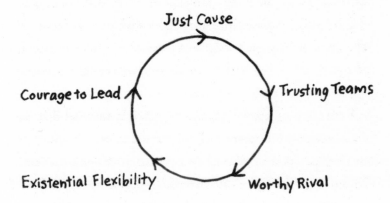

Maintaining an infinite mindset is hard. Very hard. It is to be expected that we will stray from the path. We are human and we are fallible. We are subject to bouts of greed, fear, ambition, ignorance, external pressure, competing interests, ego . . . the list goes on. To complicate matters further, finite games are seductive; they can be fun and exciting and sometimes even addictive. Just like gambling, every win, every goal hit releases a shot of dopamine in our bodies, encouraging us to play the same way again. To try to win again. We must be strong to resist that urge.

We cannot expect that we or *every* leader will lead with a perfectly infinite mindset, or that *any* leader with an infinite mindset will be able to maintain that mindset at all times. Just as it is easier to focus on a fixed, finite goal than an infinite vision of the future, it is easier to lead a company with a finite mindset, especially during times of struggle or downturn. Indeed, every one of the examples I cite in this chapter, including the affirmative examples, has, at some point in their history, been led by someone who abandoned the infinite foundation upon which the company was built to focus on more finite pursuits. In fact, finite-mindedness nearly destroyed all of these companies. Only the lucky ones that were rescued by an infinite-minded leader have gone on to become even stronger versions of themselves, more inspiring for the people who work there and more appealing to the people who buy their products.

Regardless of how we choose to play, it is essential that we be honest with ourselves and others about our choice—for our choice makes ripples. Only when those around us—our colleagues, customers and investors—know how we have chosen to play can they adjust their expectations and behaviors accordingly. Only when they know the mindset we have adopted can they figure out the short- and long-term implications for themselves. They are entitled to know how we will play so that they may make smarter decisions about who they want to work for, buy from or invest in. When they see that we have embraced the five practices of an infinite-minded leader, they can be confident that we are focused on where we are going and committed to taking care of each other along the way. They can also be confident that we will strive to resist short-term temptations and act ethically as we build our organizations to survive and thrive for a very, very long time to come.

As for us, those who choose to embrace an infinite mind-set, our journey is one that will lead us to feel inspired every morning, safe when we are at work and fulfilled at the end of each day. And when it is our time to leave the game, we will look back at our lives and our careers and say, "I lived a life worth living." And more important, when imagining what the future holds, we will see how many people we've inspired to carry on the journey without us.

Chapter 2

JUST CAUSE

First they ate the animals in the zoo. Then they ate their cats and dogs. Some even resorted to eating wallpaper paste and boiled leather. Then the unthinkable. "A child died, he was just three years old," wrote Daniil Granin, one of the survivors. "His mother laid the body inside the double-glazed window and sliced off a piece of him every day to feed her second child."

These were some of the extremes the people of Leningrad were driven to during the Nazis' nearly nine-hundred-day siege of the city from September 1941 to January 1944. Over a million citizens, including four hundred thousand children, died, many of them due to starvation. And all the while, unbeknownst to the masses, a stash of hundreds of thousands of seeds and tons of potatoes, rice, nuts and cereal lay hidden in the heart of the city.

About twenty-five years before the siege began, a young botanist named Nikolai Vavilov started building his seed collection. Growing up in a time when Russia was ravaged by major famines that killed millions of people, he committed his life and his work to ending hunger and preventing future ecological disasters. What started as idealism eventually became a highly focused cause for Vavilov. He traveled the world to collect various types of food crops and learn more about what made some more resilient than others. Before long he had collected seeds from over six thousand types of crops. He also started to study genetics and experimented with developing new strains of crops that could better resist pests or disease, grow more quickly, withstand harsh conditions or offer higher yields of food. As his work advanced, Vavilov's vision for a seed bank crystallized. Just as we keep a backup of important data should our computer crash, Vavilov wanted to have a backup of the seeds for all the world's food should any species become extinct or ungrowable due to natural or man-made disasters.

Having built up quite a reputation (and an even larger seed collection), in 1920 Vavilov left his life as an academic to become the head of the Department of Applied Botany in Leningrad. With the help of government funding, Vavilov was able to bring together a whole team of scientists to join him in his work and help advance his cause. Upon his arrival at the institution, Vavilov wrote, "I would like the Department to be a necessary institution, as useful to everybody as possible. I'd like to gather the varietal diversity from all over the world, [organize them all and] turn the Department into the treasury of all crops and other floras." And like any good visionary with an infinite mindset, he concluded, "The outcome is uncertain. . . . But still, I want to try."

Within two years, however, things had changed. This was Joseph Stalin's Soviet Union, and no one was safe. Not even the highly respected Vavilov. Over the course of his rule, which lasted from 1922 until his death in 1953, Stalin is said to have been responsible for the deaths of over 20 million of his own people. And sadly, the scientist who had devoted his life to helping his country's people found himself one of Stalin's political targets. Arrested in 1940 on trumped-up charges of espionage, Vavilov was subjected to over four hundred sessions of brutal interrogation, some lasting thirteen hours, all with the intent to break his spirit and coerce a confession that he was an anti-Stalin sympathizer. But Vavilov was not a man who could be easily broken, not even under such extreme conditions. Despite his captors' best efforts, Vavilov never broke. He never confessed to the false charges against him. Sadly, in 1943, at only fifty-five years old, the visionary botanist and plant geneticist who had devoted his life to ending hunger died in prison of malnutrition.

At the time of Vavilov's death, the siege of Leningrad was raging. There, in the middle of a war zone, hidden in a rather nondescript building in St. Isaac's Square, were the records of all the work Vavilov's team had done, and of course, their priceless seed collection, which now consisted of hundreds of thousands of varieties of crops. Beyond the obvious risks from shelling, the collection was also threatened by an explosion of rats in the city (the starving people had eaten all the cats, which would ordinarily control the rat population). And as if that weren't enough, Vavilov's collection had also caught the attention of the Nazis. Obsessed with eugenics and his own health, Hitler knew the value of the seed bank and wanted it for himself and for Germany. The problem was, although Hitler knew of its existence, he did not know its location. So he tasked a group within his army to find it.

Despite the threats, and despite being subjected to the same grueling conditions as all the other residents of Leningrad, Vavilov's team of scientists continued their work throughout the siege. They ventured out in the middle of winter, for instance, to resow secret plots of potatoes in a field near the front lines. Though they were able to smuggle some of their work out of the city, the rest they kept hidden and under guard. The scientists were so devoted to Vavilov's vision that they were prepared to protect the seed bank at any cost. Even if the cost was their lives. In the end, surrounded by hundreds of thousands of seeds, tons of potatoes, rice, nuts, cereals and other crops that they refused to eat, nine of the scientists died of starvation.

When talking about his cause, Vavilov was once quoted as saying, "We shall go into the pyre, we shall burn, but we shall not retreat from our convictions." And those who joined him in common cause were more than inspired by Vavilov's words. They lived them. One of the survivors, Vadim Lekhnovich, who helped plant the seed potatoes and stood guard over them while shots flew through the air, was later asked about not eating the bounty. "It was hard to walk. It was unbearably hard to get up every morning, to move your hands and feet," he said, "but it was not in the least difficult to refrain from eating up the collection. For it was *impossible* [to think of] eating it up. For what was involved was the cause of your life, the cause of your comrades' lives."

The scientists who carried on Vavilov's work during the

siege felt like they were a part of something bigger than themselves. This Just Cause, "a mission for all humanity," as Vavilov called it, gave their work and their lives purpose and meaning beyond any one individual or the very real struggles they faced in the moment of the siege. To have fed themselves or even to have fed the masses of starving residents in the city would have been a finite solution to a finite problem. Though they may have helped prolong the lives of some who would likely still have died or even saved the lives of others, they were looking beyond the immediate horizon. They weren't imagining the relatively few lives they could save in Leningrad; they imagined a future state in which their work might save entire civilizations. Their work was not devoted to getting to the end of the siege; they were playing to keep the human race going for as long as possible.

What a Just Cause Is

Howard's Little League team was one of the, if not the, worst in the league. At the end of each lost game, his coach would say to the players, "It doesn't matter who wins or loses, what matters is how we played the game." At which point, the precocious young Howard would raise his hand and ask the coach, "Then why do we keep score?"

When we play in a finite game, we play the game to win. Even if we hope to simply play well and enjoy the game, we do not play to lose. The motivation to play in an infinite game is completely different—the goal is not to win, but to keep playing. It is to advance something bigger than ourselves or our organizations. And any leader who wishes to lead in the Infinite Game must have a crystal clear Just Cause.

A Just Cause is a specific vision of a future state that does not yet exist; a future state so appealing that people are willing to make sacrifices in order to help advance toward that vision. Like Vavilov's scientists, the sacrifice people are willing to make may be their lives. But it needn't be. It can be the choice to turn down a better-paying job in order to keep working for an organization that is working to advance a Just Cause in which we believe. It may mean working late hours or taking frequent business trips. Though we may not like the sacrifices we make, it is because of the Just Cause that they feel worth it.

"Winning" provides a temporary thrill of victory; an intense, but fleeting, boost to our self-confidence. None of us is able to hold on to the incredible feeling of accomplishment for that target we hit, promotion we earned or tournament we won a year ago. Those feelings have passed. To get that feeling again, we need to try to win again. However, when there is a Just Cause, a reason to come to work that is bigger than any particular win, our days take on more meaning and feel more fulfilling. Feelings that carry on week after week, month after month, year after year. In an organization that is only driven by the finite, we may like our jobs some days, but we will likely never *love* our jobs. If we work for an organization with a Just Cause, we may like our jobs some days, but we will always love our jobs. As with our kids, we may like them some days and not others, but we love them every day.

A Just Cause is not the same as our WHY. A WHY comes from the past. It is an origin story. It is a statement of who we are—the sum total of our values and beliefs. A Just Cause is about the future. It defines where we are going. It describes the world we hope to live in and will commit to help build. Everyone has their own WHY (and everyone can know what their WHY is if they choose to uncover it). But we do not have to have our own Just Cause, we can choose to join someone else's. Indeed we can start a movement, or we can choose to join one and make it our own. Unlike a WHY, of which there can be only one, we can work to advance more than one Just Cause. Our WHY is fixed and it cannot be changed. In contrast, because a Just Cause is about something as of yet unbuilt, we do not know exactly the form it will take. We can work tirelessly to build it however we want and make constant improvements along the way.

Think of the WHY like the foundation of a house, it is the starting point. It gives whatever we build upon it strength and permanence. Our Just Cause is the ideal vision of the house we hope to build. We can work a lifetime to build it and still we will not be finished. However, the results of our work help give the house form. As it moves from our imagination to reality it inspires more people to join the Cause and continue the work . . . forever. For example, my WHY is to inspire people to do what inspires them so that together we can each change our world for the better. It is uniquely mine. My Just Cause is to build a world in which the vast majority of people wake up inspired, feel safe at work and return home fulfilled at the end of the day, and I am looking for as many people as possible who will join me in this Cause.

It is the Just Cause that we are working to advance that gives our work and our lives meaning. A Just Cause inspires us to stay focused beyond the finite rewards and individual wins. The Just Cause provides the context for all the finite games we must play along the way. A Just Cause is what inspires us to want to keep playing. Whether in science, nation building or business, leaders who want us to join them in their infinite pursuit must offer us, in clear terms, an affirmative and tangible vision of the ideal future state they imagine.

When the Founding Fathers of the United States declared independence from Great Britain, for example, they knew that such a radical act would require a statement of Just Cause. "We hold these truths to be self-evident, that all men are created equal," they wrote in the Declaration of Independence, "that they are endowed by their Creator with certain unalienable Rights, that among these are Life, Liberty and the pursuit of Happiness." The vision they set forth was not simply one of a nation defined by borders but of an ideal future state defined by principles of liberty and equality for all. And on July 4, 1776, the fifty-six men who signed on to that vision agreed to "mutually pledge to each other our Lives, our Fortunes and our sacred Honor." This was how much it mattered to them. They were willing to give up their own finite lives and interests to carry forward the infinite idea and ideals of a new nation. Their sacrifice, in turn, inspired subsequent generations to embrace the same Cause and devote their own blood, sweat and tears to continue to advance it.

We know a Cause is just when we commit to it with the confidence that others will carry on our legacy. This was certainly the case for America's founders. And it was the case for Nikolai Vavilov. Vavilov's vision of a world in which entire populations, and indeed all of humanity, will always have a source of food, ensuring that we can survive as long as possible, carries on to this day. There are nearly two thousand seed banks spread across more than one hundred countries around the world that are continuing the work that Vavilov started a lifetime ago. The Svalbard Global Seed Vault in Norway is one of the largest. Located in a naturally temperature-controlled environment in the Arctic, the Svalbard Vault stores over a billion seeds from nearly six thousand species of flora. It is there to ensure that in the worst-case scenario, we would have a food source to keep our species alive. Marie Haga, the executive director of the Crop Trust, the organization formed in partnership with the United Nations to help support the work of global seed banks, points to Vavilov as the ostensible founder of the cause. "A century after [Vavilov's] first journeys," she said, "a new generation of dedicated crop diversity supporters continue to travel the world to conserve not only germplasm but also Vavilov's legacy."

Many of the organizations we work for now already have some sort of purpose, vision or mission statement (or all of them) written on the walls that our leaders hope will inspire us. However, the vast majority of them would not qualify as a Just Cause. At best they are uninspiring and innocuous, at worst they point us in a direction to keep playing in the finite realm. Even some of the best-intentioned attempts are written in a way that is finite, generic, self-centered or too vague to be of any use in the Infinite Game. Common attempts include statements like, "We do the stuff you don't want to do, so that you can focus on the things that you love to do." It may be a true statement, it's just a true statement for too many things, especially in a business-to-business space. Plus, it's not much of a rallying cry. Another common generic vision sounds like, "To offer the highest quality products at the best possible value, etc., etc." Statements like this are of little use for those who wish to lead us in the Infinite Game. Such statements are not inclusive. They are egocentric—about the company; they look inward and are not about the future state to which the products or services are contributing.

Vizio, the California-based maker of televisions and speakers, says on their website, for example, that they exist to "deliver high performance, smarter products with the latest innovations at a significant savings that we can pass along to our consumers." I take them at their word that they do all those things. But do those words *really* inspire people to want to offer their blood, sweat or tears? When you read those words are you inspired to rush to apply for a job there? Few if any of us get goose bumps or feel a visceral calling to be a part of something like that. Such statements offer us neither a cause to which we would commit ourselves nor a sense of what it's all for, both of which are essential in the Infinite Game.

Again, a Just Cause is a specific vision of a future state that does not yet exist. And in order for a Just Cause to provide direction for our work, to inspire us to sacrifice, and to endure not just in the present but for lifetimes beyond our own, it must meet five standards. Those who are unsure whether their purpose, mission or vision statement is a Just Cause or those interested in leading with a Just Cause can use these standards as a simple test.

A Just Cause must be:

- **For something**—affirmative and optimistic
- **Inclusive**—open to all those who would like to contribute
- **Service oriented**—for the primary benefit of others
- **Resilient**—able to endure political, technological and cultural change
- **Idealistic**—big, bold and ultimately unachievable

For something—affirmative and optimistic

A Just Cause is something we stand for and believe in, not something we oppose. Leaders can rally people *against* something quite easily. They can whip them into a frenzy, even.

For our emotions can run hot when we are angry or afraid. Being *for* something, in contrast, is about feeling inspired. Being *for* ignites the human spirit and fills us with hope and optimism. Being *against* is about vilifying, demonizing or rejecting. Being *for* is about inviting all to join in common cause. Being *against* focuses our attention on the things we can see in order to elicit reactions. Being *for* focuses our attention on the unbuilt future in order to spark our imaginations.

Imagine if instead of fighting *against* poverty, for example, we fought *for* the right of every human to provide for their own family. The first creates a common enemy, something we are against. It sets up the Cause as if it is "winnable," i.e., a finite game. It leads us to believe that we can defeat poverty once and for all. The second gives us a cause to advance. The impact of the two perspectives is more than semantics. It affects how we view the problem/vision that affects our ideas on how we can contribute. Where the first offers us a problem to solve, the second offers a vision of possibility, dignity and empowerment. We are not inspired to "reduce" poverty, we are inspired to "grow" the number of people who are able to provide for themselves and their families. Being for or being against is a subtle but profound difference that the writers of the Declaration of Independence intuitively understood.

Those who led America toward independence stood *against* Great Britain in the short term. Indeed the American colonists were deeply offended by how they were treated by England. Over 60 percent of the Declaration of Independence is spent laying out specific grievances against the king. However, the Cause they were fighting *for* was the true source of lasting inspiration, and in the Declaration of Independence it came before anything else. It is the first idea we read in the document. It sets the context for the rest of the Declaration and the direction for moving forward. It is the ideal to which we personally relate and that we have easily committed to memory. Few Americans, except for scholars and the most zealous of history buffs, can rattle off even one of the complaints listed later in the document, things like: "He has endeavored to prevent the Population of these States; for that purpose obstructing the Laws for naturalization of foreigners; refusing to pass others to encourage their Migrations hither, and raising the Conditions of new Appropriations of Lands." In contrast, most Americans can recite with ease "all men are created equal" and can usually rattle off the three tenets of "Life, Liberty and the pursuit of Happiness." These words are indelibly marked on the cultural psyche. Invoked by patriots and politicians alike, they remind Americans of who we strive to be and the ideals upon which our nation was founded. They tell us what we stand *for*.

Inclusive—open to all those who would like to contribute

Human beings want to feel a part of something. We crave the feeling of belonging. We enjoy the feeling of being part of a group, like when we attend church, attend a parade or rally or wear the jersey of our favorite team when we attend a sports event. A Just Cause serves as an invitation to join others in advancing a cause bigger than ourselves. When the words of the Just Cause help us imagine a positive, specific, alternative vision of the future, it stirs something inside us that makes us want to raise our hand to join up and join in.

A well-crafted statement of Cause inspires us to offer our ideas, our time, our experience, our hands, anything that may help advance the new vision of the future it articulates. This is how movements come to be. It starts with a few people. Their idealized vision of the future attracts believers. Those early adopters don't show up to get anything, they show up to give. They want to help. They want to play a role in advancing toward a new version of the future. The Cause that attracted them becomes their own.

Organizations that simply promise to "change the world" or "make an impact" tell us very little about what specifically they want to accomplish. The sentiments are good, but they are too generic to serve as a meaningful filter for us. Again, a Just Cause is a *specific* vision of a future state that does not yet exist; a future state so appealing that people are willing to make sacrifices in order to help advance toward that vision. We call it "vision" because it must be something we can "see." For a Just Cause to serve as an effective invitation, the words must paint a specific and tangible picture of the kind of impact we will make or what exactly a better world would look like. Only when we can imagine in our mind's eye the exact version of the world an organization or leader hopes to advance toward will we know to which organization or to which leader we want to commit our energies and ourselves. A clear Cause is what ignites our passions.

"We only hire passionate people" is the oft-recited standard of many a person responsible for hiring. How do they know, however, whether the candidate is passionate for interviewing but not so passionate for the Cause? The reality is, EVERYONE is passionate about something, but we aren't all passionate about the same thing. Infinite-minded leaders actively seek out employees, customers and investors who share a passion for the Just Cause. For employees, this is what we mean when we say, "Hire for culture and you can always teach the skills later." For customers and investors, this the root of love and loyalty for the organization itself.

The quick-serve salad company Sweetgreen stands for something bigger than selling salads, for example, and they invite would-be contributors to join their Cause. Their stated mission is "to inspire healthier communities by connecting people to real food." Real food, as Sweetgreen defines it, means ingredients from local sources that support local farms. Which is why their stores have different menus depending on which part of the country they are in. Though many of us may buy their salads just because we like their salads, those who are devoted to locally sourced food and want to support local farms will be drawn to work for and become the most loyal supporters of Sweetgreen. They will make sacrifices, like going out of their way or paying a premium, to buy from Sweetgreen. Supporting the company in some shape or form is one of the things they do to feel that they are advancing their own values and beliefs, their own vision of a better world. They feel included in the Cause.

Service oriented—for the primary benefit of others

A Just Cause must involve at least two parties—the contributors and the beneficiaries. The givers and receivers. Contributors give something, e.g., their ideas, hard work or money, to help advance the Just Cause. And the receivers of those contributions benefit. For a Just Cause to pass the service-orientation test, the primary benefit of the organization's contributions must always go to people other than the contributors themselves.

If my boss offers me career advice, for example, that advice must be for the primary benefit of my career and not theirs. If I am an investor, I must intend that the primary benefit of my contribution goes to helping the company advance its Just Cause. If I am a leader, I must intend that the primary benefit of my time, effort and decisions goes to those I lead. If I am a frontline employee, I must intend that the primary benefit of my efforts goes to the people buying our product or service. If there is only one party, if we are the sole beneficiaries of our work, that's not a Just Cause, that's a vanity project.

When Sweetgreen talks about the beneficiaries of its contributions, they talk about communities and people. They don't talk about what their contributions will do for Sweetgreen. And the drafters of the Declaration of Independence were clear that "We the people," not "We the leaders," would be the primary beneficiaries of their efforts and of the Revolution. If those who led the fight had made themselves the primary beneficiaries, then America probably would have ended up with a dictatorship or an oligarchy. With that new perspective, we instantly see what follows when a company says the primary beneficiaries of their work are shareholders, not customers.

The operative word in all this is "primary." Service orientation does not mean charity. In charity, the vast majority, if not all, the benefit of our contributions must go to the receiver. And any benefit the contributor gets is the good feeling that they contributed. In business, of course we can consider how our work will benefit us or advance our own lot. Of course we can expect and even demand to be fairly compensated and recognized for our efforts and results. We can want our investors to benefit too, just not at the expense of the company, the people who work there or the customers who buy from us. No beneficiary, no customer, should be forced to buy a substandard product and no employee should lose their job as a result of cost cutting performed to benefit a shareholder, who is, after all, just one of a group of contributors. Again, only when the primary beneficiary of the Cause is someone other than the organization itself can the Cause be Just.

This is what "servant leadership" means. It means the primary benefit of the contributions flows downstream. In an organization where service orientation is lacking (or treated as a sideshow rather than the main event), the flow of benefits tends to go upstream instead. Investors invest with the primary intention of seeing a return before anyone else. Leaders make decisions that benefit themselves before those in their charge. Salespeople ensure they do whatever they need to do to make the sale to earn their bonus, regardless of what the customer needs. This is the common flow of benefit in so many of our organizations today. Too many of our cultures are filled with people working to protect their own interests and the interests of those above them before those of the people they are supposed to be serving.

The requirement that a Just Cause be service oriented is consistent with how infinite games are supposed to be played. The infinite player wants to keep the game going for others. A leader who wishes to build an organization equipped for the Infinite Game must never make decisions solely to boost their own compensation. Their efforts should go toward equipping the organization for the game in which it is operating. Even an investor must not be the primary beneficiary of their investment. Rather it is the organization in which they believe and whose Just Cause they want to see advanced that must benefit from their financial contribution. An infinite-minded investor wants to contribute to advance something bigger than themselves—which, if it is successful, will be highly profitable. A finite-minded investor is more like a gambler who bets solely so they may reap the reward. Let us not confuse the two behaviors.

The reason a service orientation is so important in the Infinite Game is because it builds a loyal base of employees and customers (and investors) who will stick with the organization through thick and thin. It is this strong base of loyalty that gives any organization a kind of strength and longevity that money alone cannot provide. The most loyal employees feel their leaders genuinely care about them . . . because their leaders genuinely do care about them. In return, they offer their best ideas, act freely and responsibly and work to solve problems for the benefit of the company. The most loyal customers feel the company genuinely cares about their wants, needs and desires . . . because the company really does. And in return, this is why loyal customers go out of their way or pay a premium to buy from that company over another and encourage their friends to do the same. And the best-led companies feel like their investors genuinely care about helping the company become as strong as possible in order to advance the Cause because the investors really do care. The results benefit all stakeholders.

Resilient—able to endure political, technological and cultural change

Leaders who wish to lead with an infinite mindset would do well to keep the example of the Declaration of Independence in mind. The founders' stated commitment to equality and unalienable human rights are evergreen. Over the course of more than 240 years, even as the nation's leaders, landscape, people and culture have changed, the Just Cause has remained as relevant and inspiring as ever. It is a Just Cause for an infinite time frame.

In the Infinite Game of business, a Just Cause must be greater than the products we make and the services we offer. Our products and services are some of the things we use to advance our Cause. They are not themselves the Cause. If we articulate our Cause in terms of our products, then our organization's entire existence is conditional on the relevance of those products. Any new technology could render our products, our Cause and indeed our entire company obsolete overnight. The American railroads, for example, were some of the largest companies in the country. Until advancements in automotive technology and a network of highways offered people a quicker and sometimes cheaper alternative to the train. Had the railroads defined their need to exist in terms related to moving people and things instead of advancing the railroad, they might be the owners of major car companies or airlines today. Publishers saw themselves in the book business instead of the spreading-ideas business and thus missed the opportunity to capitalize on new technology to advance their cause. They could have invented Amazon or the digital e-reader. Had the music industry defined themselves as the sharers of music rather than sellers of records, tapes and CDs they would have had an easier time in a world of digital streaming. By defining themselves by a cause greater than the products they sold, they could have invented services like iTunes or Spotify. But they didn't . . . and now they are paying the price for it.

Markets will rise and fall, people will come and go, technologies will evolve, products and services will adapt to consumer tastes and market demands. We need something with permanence for us to rally around. Something that can withstand change and crisis. To keep us in the Infinite Game, our Cause must be durable, resilient and timeless.

Idealistic—big, bold and ultimately unachievable

When the signers of the Declaration of Independence affirmed that all men "are created equal" and "endowed . . . with certain unalienable Rights," they were referring primarily to white, Anglo-Saxon, Protestant men. Almost immediately, however, there were efforts to advance a more expansive and inclusive understanding of the ideal. During the Revolutionary War, for example, George Washington forbade anti-Catholic organizing in his armies and regularly attended Catholic services to model the behavior he expected of his men. Nearly a hundred years later, the Civil War brought about an end to slavery, and soon after that the Fourteenth Amendment granted citizenship and equal rights to African Americans and former slaves. The women's suffrage movement took another step toward America's Just Cause when it gained the vote for women in 1920. The Civil Rights Act of 1964 and the Voting Rights Act of 1965, which protected African Americans and others from discrimination, were two more steps. The nation took yet another step in 2015 with the Supreme Court decision in *Obergefell v. Hodges*, which extended the protections guaranteed by the Fourteenth Amendment to gay marriage.

If the founders of the United States had only set out a goal—to win independence—once it was achieved, they would have grabbed a pint of ale and sat around playing rounds of ninepins and ring taw while regaling each other with how great it was that they won the war. But that's not what happened. Instead, they got to work writing a constitution (which was only fully ratified seven years after the official end of the Revolutionary War) to further codify a set of enduring principles to protect and advance their big, bold, idealistic vision of the future. A vision that Americans have been striving to protect and advance ever since quill and ink touched paper . . . and will continue to protect and advance as long as we have the will and resources to do so. America's Just Cause has yet to be fully realized, and for all practical purposes it never will be. But we will die trying. And that's the point.

Indeed, the abolition of slavery, women's suffrage, the Civil Rights Act and gay rights are some of the big steps the nation has taken to realize its Cause. And though each of those movements, infinite in their own right, are still far from complete, they still represent clear steps along the nation's march toward the ideals enshrined in the Declaration of Independence. It is important to celebrate our victories, but we cannot linger on them. For the Infinite Game is still going and there is still much work to be done. Those victories must serve as milestones of our progress toward an idealized future. They give us a glimpse of what our idealized future can look like and serve as an inspiration to keep moving forward.

This is what the idealized journey of a Just Cause feels like—no matter how much we have achieved, we always feel we have further to go. Think of a Just Cause like an iceberg. All we ever see is the tip of that iceberg, the things we have already accomplished. In an organization, it is often the founders and early contributors who have the clearest vision of the unknown future, of what, to everyone else, remains unseen. The clearer the words of the Just Cause, the more likely they will attract and invite the innovators and early adopters, those willing to take the first risks to advance something that exists almost entirely in their imaginations. With each success, a little more of the iceberg is revealed to others; the vision becomes more visible to others. And when others can see a vision become something real, skeptics become believers and even more people feel inspired by the possibility and willingly commit their time and energy, ideas and talents to help advance the Cause further. But no matter how much of the iceberg we can see, our leaders have the responsibility to remind us that the vast majority still lies unexplored. For no matter how much success we may enjoy, the Just Cause for which we are working lies ahead and not behind.

When You Have Your Cause, Write It Down

The Founding Fathers of the United States were larger-than-life figures. They lived and breathed their Just Cause. This is often the case with inspirational leaders in business as well. But what happens when those charismatic keepers of the Cause move on, retire or die? I am often surprised how many visionary leaders don't think they need to find the words for or write down their Cause. They assume that because their vision is clear to them it's clear to everyone else in the organization. Which of course it's not.

Without finding the words for the Just Cause and writing them down, it dramatically increases the risk that, in time, the Cause will be diluted or disappear altogether. And without the Just Cause, an organization starts to function like a ship without a compass—it veers off course. Focus moves from beyond the horizon to the dials in front of them. Without a Just Cause to guide them, finite-mindedness starts to creep in. The leaders will celebrate how fast they are going or how many miles they have traveled, but fail to recognize that their journey lacks any direction or purpose.

A Just Cause that is preserved on paper can be handed down from generation to generation; a founder's instinct cannot. Like the Declaration of Independence, a written statement of Cause dramatically increases the chances that the Cause will survive to guide and inspire future generations beyond the founders and those who knew the founders. It's the difference between a verbal contract and a written contract. Both are legal and enforceable, but when a contract is written it prevents any confusion or disagreement about the terms of the deal . . . especially for people who weren't there when the deal was made.

A written cause works like a compass. And with a compass in hand, each succession of leaders, their gaze looking beyond the horizon, can more easily navigate the technologies, politics and cultural norms of the day without the founder present.

Chapter 3

CAUSE. NO CAUSE.

Let's play a quick round of Cause. No Cause.

It is a good thing that more and more companies seem to be embracing the importance of having a purpose at the heart of their business. The problem is, too many of them say things that only sound like a Just Cause. Indeed, they may even use language and meet some of the standards of a Just Cause. Until they can check all five boxes, however, what they offer simply isn't a Just Cause.

There are a few main reasons we fail to put forward a true Just Cause. Sometimes, the visionary, Cause-driven leader adopts a false cause by accident because they are struggling to find the words to embody what they imagine for the future (see previous chapter for help). In other cases, the leader wants people to believe that they are Cause driven when, in fact, they have no vision at all. Common "imposter causes" include things like moon shots, a drive to "be the best," or mistaking "growth" for purpose. It is also common to find organizations confusing their corporate social responsibility (CSR) program for a Just Cause. Any of these may or may not work in the finite game, but they absolutely cannot lead an organization to survive and thrive in the Infinite Game.

The reason to identify these pitfalls is first, as a warning, that to embrace any of these will not prepare an organization for life in the Infinite Game, but rather keep it playing squarely with a finite mindset. The other reason to point them out is simply so that we can know if indeed we have a Just Cause or not and go back to the drawing board if we need to. We might even avoid coming up with a false cause in the first place. An organization that has a false cause is not a bad company, it just means they may have a little more work to do. The ability to recognize false causes can also save us pain as investors, employees and consumers. If we suspect that one organization does not have a Just Cause, we can move on to another that does.

A true Just Cause is deeply personal to those who hear it, and it must be deeply personal to those who espouse it. The more personal it is for people, the more likely our passions will be stoked to help advance it. If the words of a Just Cause are used simply to boost a brand image, attract passionate employees or help drive some near-term goal, like a purchase, a vote or support for the company, the impact will be short lived. As soon as we start working at an organization or interacting with its people, we will quickly find out whether they are offering us a Just Cause they truly believe in or just hollow words.

Moon Shots Are Not a Just Cause

He offered us something to believe in. Something that was bigger than us. Something that we were willing to sacrifice to see happen. "We choose to go to the moon," said President John F. Kennedy with determination. "We choose to go to the moon in this decade . . . not because [it is] easy, but because [it is] hard, because that goal will serve to organize and measure the best of our energies and skills, because that challenge is one that we are willing to accept, one we are unwilling to postpone, and one which we intend to win." And just over eight years after Kennedy first challenged the nation, Neil Armstrong took "one small step for a man, one giant leap for mankind."

The so-called moon shot is often invoked by leaders who are trying to inspire their people to reach for something that seems impossible. And because moon shots pass most of the tests of a Just Cause, it usually works. In the case of Kennedy's actual moon shot, it is affirmative and specific. It is inclusive, service oriented and definitely worthy of sacrifice. However, it is not infinite. No matter how hard the challenge, no matter how impossible it seemed, the moon shot was an achievable, finite goal. More than an ideal future state, it is what Jim Collins, author of *Good to Great* and *Built to Last,* calls a BHAG, a big, hairy, audacious goal. It's easy to mistake a BHAG for a Just Cause because they can indeed be incredibly inspiring and can often take many years to achieve. But after the moon shot has been achieved the game continues. Simply choosing another big, audacious goal is not infinite play, it's just another finite pursuit.

During employee town hall meetings at GE, some of the employees would express concern that the company was too

focused on the short term. Jack Welch, then CEO, was fond of replying, "Long term is just a series of short terms." When employees express such a concern to a CEO, more likely than not what they are really asking is: "What's this all for?" What is all our hard work contributing to beyond the metrics and material rewards? Welch's answer revealed that, to him, there was no higher cause at play. The goal was simply to perform, perform again and perform again. To Welch, each finite accomplishment was enough. Except, business is an infinite game, which means the series of short terms never ends.

Indeed, leaping from goal to goal can be fun for a while, but if that's all there is, over time the thrill of each achievement becomes less, well, thrilling. I often meet senior executives who seem to suffer from a kind of "finite exhaustion." Because they did well and were paid well for hitting each goal set for them, they kept repeating that pattern. At some point in their careers, they traded any fantasy of feeling like their work would contribute to something bigger than themselves for a rat race or a hamster wheel or some other unfulfilling running rodent metaphor. Racking up finite wins does not lead to something more infinite.

The question that a Just Cause must answer is: What is the infinite and lasting vision that a moon shot will help advance? A Just Cause is the context for all our other goals, big and small, and all of our finite achievements must help to advance the Just Cause. Indeed, if we become overly concerned with a finite goal, no matter how inspiring, we leave ourselves open to making decisions that are only good for the finite but may do damage to the infinite.

Kennedy's moon shot was made in the context of the larger infinite vision that America's Founding Fathers laid out—that our progress is not for the benefit of a few, but for the benefit of many. In the sentences before Kennedy proposed his moon shot challenge, he offered the infinite context for the finite objective, "We set sail on this new sea because there is new knowledge to be gained, and new rights to be won, and they must be won and used for the progress of all people." This was his belief for a good many of his objectives, including landing a man on the moon and returning him home safely.

Though moon shots are inspiring for a time, that inspiration comes with an expiration date. Moon shots are bold, inspiring finite goals *within* the Infinite Game, not *instead* of the Infinite Game.

Being the Best Is Not a Just Cause

"We will be the global leader in every market we serve and our products will be sought after for their compelling design, superior quality, and best value." This is a pretty typical-sounding corporate vision or mission statement. This one belongs to Garmin, the maker of GPS devices for everyone from runners to pilots. Though there are dozens of variations, the basic formula is the same—we're the best and everyone wants our products because our products are the best . . . and "they're great value" (gotta squeeze that in).

Again, vision or mission statements act like compasses. They guide our direction. However, because there are no standards on how to write such statements, ones like the above have become too common. Broad and generic, they offer little to no value to a company that wants to adopt an infinite mindset. "Being the best" and statements like that are egocentric statements that place the company as the primary subject (and thus the primary beneficiary) of their vision. They don't help make the company relevant to those who buy from the company. In fact, any mention of the customer or any offer of value usually comes at the end of the statement. By putting the egocentric statement first, it directs leaders to focus their efforts inward and not on actual people who may buy the product. And just because people may buy or like the product does not mean they believe in or even know what the Cause is.

Leaders with a finite mindset often confuse having a successful product with having a strong company. Which is a little like the owners of the Los Angeles Lakers thinking their team is relevant because LeBron James has relevance. Having a great player, a popular product or a killer app does not mean we are equipped for the Infinite Game. Vision statements that place the product at the center of the vision are only useful so long as nothing better ever comes along, there is never a deviation in market conditions and no new technology is ever invented. If, however, any of these things does happen, the company will be left with a vision statement that often leaves them clinging on to an old business model and blind to the opportunities they could have captured. This seems to be what happened to Garmin.

In 2007, Garmin may have been "the best," the global leader in dash-mounted GPS units for cars and boats. However, as smartphones became more reliable and more capable, we had less need for a separate GPS unit anymore and the company suffered as a result. It is now worth less than a third of what it was worth in 2007. It's too easy for Garmin to simply blame the rise and ubiquity of smartphones to explain their losses (which they did). What they failed to recognize is that they had a vision statement that directed them to focus on their product, and in so doing, they missed the opportunity that smartphones offered them. Had they been obsessing about how to provide the value to customers first, they may have seized the chance to develop the go-to navigation app for mobile phones when the opportunity still existed. Their brand was certainly strong enough to do so. Instead, they continued to focus on the business model they had selling dash-mounted hardware. Now the default navigation apps on our phones are Google Maps, Waze or Apple Maps, but that didn't have to be. A Just Cause should direct the business model, not the other way around.

When a statement of vision or mission is grounded in the product, it can have adverse effects on the corporate culture also. For companies that place their product above all else, which is fairly common among technology or engineering companies, it leaves people who are not engineers or product designers feeling like (and sometimes actually treated like) second-class citizens in their own companies. An organization is better served if everyone, including those in accounting, support or customer service roles, for example, is made to feel like they are not just there to serve the needs of the engineers or product development teams. They too want to feel like valuable members of the team, working together to advance something bigger than the product or themselves.

Being the best simply cannot be a Just Cause, because even if we are the best (based on the metrics and time frames of our own choosing), the position is only temporary. The game doesn't end once we get there; it keeps going. And because the game keeps going, we often find ourselves playing defense to maintain our cherished ranking. Though saying "we are the best" may be great fodder for a rah-rah speech to rally a team, it makes for a weak foundation upon which to build an entire company. Infinite-minded leaders understand that "best" is not a permanent state. Instead, they strive to be "better." "Better" suggests a journey of constant improvement and makes us feel like we are being invited to contribute our talents and energies to make progress in that journey. "Better," in the Infinite Game, is better than "best."

Growth Is Not a Just Cause

Imagine you walk out of your house one morning and see your neighbor packing up his car. "Where are you going?" you ask. "Vacation," he replies. "Nice. Where are you going?" you follow up, curious. "I told you, vacation," he replies again. "I got that," you say, "but *where* are you going?" Exasperated, your neighbor replies again, "I told you, VAY-CAY-SHUN!"

Realizing that your line of questioning will not reveal the answer you're looking for, you try a new strategy. "Okay," you say, "how do you plan to get to your vacation?" And immediately your neighbor offers their plan. "I'm going to drive down the I-90 and my goal is to drive three hundred miles per day."

If the question asked is, "What is your company's Cause? Why does your company exist?" and the answer offered is "growth," that's a lot like your neighboring responding "vacation" to the question "Where are you going?" The leaders of these growth-oriented companies can rattle off their strategies and targets for growth, but that's like explaining which highway and how many miles you plan to travel when heading on vacation; it doesn't paint a picture of why you set off in the first place or where you hope to go. It doesn't offer a larger context or purpose for that growth.

Money is the fuel to advance a Cause, it is not a Cause

itself. The reason to grow is so that we have more fuel to advance the Cause. Just as we don't buy a car simply so we can buy more gas, so too must companies offer more value than their ability to make money. A company, like a car, is more valuable to all constituents when it takes us somewhere to which we would otherwise be unable to go. That place we envision going to is the Just Cause.

It's worth noting that so many of the goals that companies put forward tend to be arbitrary or overly ambitious. Especially in the start-up world, the drive for billion-dollar valuations is not an indicator of a healthy company that is built to last. It is a standard that has evolved thanks to the venture capital industry (because valuations are how they make their money). A strong culture and the ability to fund its own existence (also known as profitability) is how a company actually stays in the game for the long term. In addition, the constant drive for hypergrowth creates a problem within mature markets—markets in which the product, technology or business is no longer new or special, but accepted and ubiquitous. For companies in those markets, companies like Sears or GE, their options are unattractive if they maintain a growth-at-all-costs mentality. Many start to play defense, give their money away to shareholders to court their favor or over use stock buybacks to keep their stock price artificially inflated. Growth through acquisition or merger often becomes the only way mature, finite-minded companies can continue to demonstrate high rates of growth. This may win a short-term boost in the stock market; however, as *Harvard Business Review* and many others have reported, "70%–90% of acquisitions are abysmal failures."

To offer growth as a cause, growth for its own sake, is like eating just to get fat. It pushes executives to consider strategies that demonstrate growth with little to no consideration of any sense of purpose for that growth. Just like it would affect a human being, it should come as no surprise that the organizations that eat to get fat will eventually suffer from health problems. Growth as a cause often results in an unhealthy culture, one in which short-termism and selfishness reign supreme, while trust and cooperation suffer. Growth is a result, not a Cause. It's an output, not a reason for being. When we have a Just Cause, we are willing to sacrifice our interests to advance it. When we think money or growth is the Cause, we are more likely to sacrifice others or the Cause itself to protect our interests. Besides, nothing can grow forever. All balloons and bubbles eventually burst . . . even financial ones.

Corporate Social Responsibility Is Not a Just Cause

The company advertised all the good they did in the community. They shared the stories of some of the people who benefited from the scholarships they funded, for example. They wanted their customers and their employees to know they cared about people. Which would have been great if the 60,000 people who actually worked for the company didn't have to work in such a top-heavy, dog-eat-dog toxic culture.

A corporate social responsibility (CSR) program is not a Just Cause. And a company is not cause driven because they sponsor walkathons, donate to charity or give employees paid time off to volunteer. Nor are they cause driven because they give away their products to people who can't afford them.

CSR programs are, for the most part, business-speak for giving to charity. And though having a CSR program is indeed great and commendable, unless you're a charity, it's only a piece of what a company does. The CSR program must be part of the broader strategy to advance the Just Cause. A strategy that includes everything the company does. The way a company makes its money and the way it gives it away must both contribute to advancing the Just Cause. "Cause-related work" is not something an organization does on the side; it is core to their very being. Service is not an ornament. It is a touchstone. And no amount of corporate social responsibility is enough to offset or balance the excessive finite focus that may consume the rest of the corporate culture.

Even well-intended finite-minded leaders often have the perspective of "make money to do good." An infinite perspective on service, however, looks somewhat different: "Do good making money" (the order of the information matters). I will do good in how I treat people and serve my community throughout my life and still build a financially strong organization. It is not so much an equation as it is a lifestyle. These individuals and companies work to be stewards of the lives of those who work for them and for the communities in which they operate. The giving that happens during and at the end of their lives looks more like a continuation of what they've been doing for decades rather than an attempt at balancing the past. The difference is determined by the leaders' mindset.

Chapter 4

KEEPER OF THE CAUSE

S am Walton founded Walmart in 1962 with a simple idea—to serve the average workin' American by offering "the lowest prices anytime, anywhere." At the end of his life, Walton described his vision this way: "If we work together, we'll lower the cost of living for everyone . . . we'll give the world an opportunity to see what it's like to save and have a better life." With Walton at the helm, the decisions that went into building Walmart—from where to locate the stores to how big they would be—were all made with this Cause at the forefront. And as a result, people *loved* Walmart—both those who worked there and those who shopped in their stores. People wanted Walmart stores in their communities. The business grew, and Walton, who had grown up during the Depression, became one of the richest men in America.

And then, somewhere along the way, the Just Cause went fuzzy. By the time Mike Duke took over as CEO in 2009, it was clear that it was no longer the driving force behind the company. Indeed, Walton's original vision was now little more than marketing slogans and hollow words written on the office walls. The company had become obsessed with profit, growth and dominance at the expense of the very Cause that drove their success in the first place.

Mike Duke earned a reputation at Walmart for being an expert in efficiency. When it was announced that Duke would be the next CEO, his predecessor, H. Lee Scott Jr., stammered, "I kind of thought—and I think the board thought—that the company could be better managed." He went on to explain, "Mike is not only a good leader but a really good manager. . . . I don't think in business you can forget the fact that you don't just have to lead, you have to manage." If the board was hoping to correct management issues or enhance performance, then giving a man like Mike Duke the reins might have been a perfect choice . . . for the short term. But if the board was concerned that Sam Walton's Just Cause had been diluted, then a man like Mike Duke was about the worst person to get the company back on track.

Duke's own words when he accepted the position revealed the kind of mindset with which he was going to lead. "[Walmart] is very well positioned in today's economy, growing market share and returns, and is more relevant to its customers than ever," he said in the press release announcing his new role. "Our strategy is sound and our management team is extremely capable. I am confident we will continue to deliver value to our shareholders, increase opportunity for our over 2 million associates, and help our 180 million customers around the world save money and live better."

Notice the order of the information? Duke's first thought was growing market share and returns. Though he talks about being relevant to customers he doesn't actually mention delivering value to them until the end of his statement. It's a strange quirk of human nature. The order in which a person presents information more often than not reveals their actual priorities and the focus of their strategies. Where Sam Walton started with the people's interests, Mike Duke started with Wall Street's.

Under Duke's leadership, Walmart's stock price did increase . . . for a while. However, focusing on numbers before people comes at a cost. The once beloved brand also found itself embroiled in multiple scandals over the treatment of their people and their customers. In 2011, Walmart was the target of one of the largest employment discrimination class action suits ever filed, brought by female employees who claimed they were victims of systematic underpayment and underpromotion. In 2012, there were walkouts and protests by workers who demanded to be treated with dignity and respect and paid a livable wage. Where before communities would rally to bring a Walmart into their neighborhoods, now they were rallying to keep them out. The company's plans for expansion in Denver and New York, for example, were halted by mass protests. There was also a congressional investigation into allegations that Walmart bribed foreign officials to court favor abroad. Needless to say, morale at the company plummeted and much of the love people had for the stores was replaced with contempt.

What happened at Walmart happens all too often in public companies, even the Cause-driven ones. Under pressure from Wall Street, we too often put finite-minded executives in the highest leadership position when what we actually need is a visionary, infinite-minded leader. Steve Ballmer, as we've already discussed, was one such example. John Sculley, who replaced Steve Jobs at Apple in 1983, was another. Instead of trying to continue advancing the Cause, Sculley was more focused on competing head-to-head against IBM. The damage he did to the culture seriously hurt Apple's ability to innovate. In 2000, after being passed over for the CEO job at GE, Robert Nardelli took over at Home Depot (his nickname at GE was "Little Jack," because of how much he emulated and hoped to succeed Jack Welch as CEO). His relentless drive for cost cutting all but destroyed a culture of innovation at Home Depot. In 2004, the COO, Kevin Rollins, replaced Michael Dell to become CEO of Dell. Focused on growth, he presided over the largest layoffs in the company history, a rise in customer complaints and an SEC investigation over accounting issues. These men were all skilled executives. However, their finite mindsets left them ill qualified for the job they had been given. In fact, Sculley at Apple and Rollins at Dell did such damage to their respective organizations that their more infinite-minded predecessors, Steve Jobs and Michael Dell, were brought back to try to repair the messes they made. The problem isn't how skilled an executive is when they take over as CEO. The problem is whether they have the right mindset for the job they are given.

We Need a New Title

The responsibility of every C-level executive is baked into their title. Chief FINANCIAL Officer. Chief MARKETING Officer. Chief TECHNOLOGY Officer. Chief OPERATING Officer. What they are required to do, what they are required to oversee, is right there in their title. One of the things that title does is to help ensure that we put the right person in the right job. Few would ever consider someone who hates numbers and has never been able to understand a balance sheet for a CFO position. And if you find technology confusing and still have that old VCR connected to your TV at home, odds are you're not on any short list to be a CTO anytime soon. So it begs the question, what exactly is a Chief EXECUTIVE Officer?

The lack of a clear standard for the role and responsibilities of the CEO in our organizations is one of the reasons we find too many leaders of companies playing the finite game when they should at least be thinking about the Infinite Game. In too many cases, it's that their title hasn't properly set them up for the job they have. The word "executive" doesn't tell us what a CEO is responsible for.

Words matter. They give direction and meaning to things. Pick the wrong words, intentions change and things won't necessarily go as hoped or expected. Martin Luther King Jr. gave the "I have a dream" speech, for example. He didn't give the "I have a plan" speech. There is no doubt he needed a plan. We know he had meetings to discuss the plan. But as the "CEO" of the civil rights movement, Dr. King was not responsible for making the plan. He was responsible for the dream and making sure those responsible for the plans were working to advance the dream.

General Lori Robinson, who, when she retired from the Air Force in 2018, was the highest-ranking female officer in the history of the United States military, explains that the responsibility of the most senior person in an organization is to look beyond the organization. "I will go up and out. I need you to go down and in" is how she framed her responsibility every time she took a new command. If the top person needs to focus on "up and out," then we need their title to help frame their primary responsibility.

Leaders in the Infinite Game will be better equipped to fulfill their responsibilities if they understand that they are stepping into the role of a "Chief Vision Officer," or CVO. That is the primary job of the person who sits at the pointy end of the spear. They are the holder, communicator and protector of the vision. Their job is to ensure that all clearly understand the Just Cause and that all other C-level executives direct their efforts to advancing the Cause inside the organization. It's not that an infinite-minded leader is entirely unconcerned with the organization's finite interests. Rather, as the keeper of the Cause, they take accountability for deciding when short-term finite costs are worth it to advance the infinite vision. They think beyond the bottom line. As the ultimate infinite player, the CVO must go up and out.

Next in Line for the Top Job

In too many of our companies today, we organize around a single line of hierarchy. The CEO is the number one job and CFO or COO are usually seen as number two. And in the vast majority of businesses, most CFOs or COOs see themselves in line for the "top job." Michael Dinkins, who worked at GE for 17 years under Jack Welch, explained:

> I think one of the reasons why a lot of CFOs are being elevated to the CEO role is because the CFO is one of the few positions that sees the total company. Everything that's going on within the company. . . . They understand processes within the company and the time frame of these processes to happen. . . . They see

how HR is recruiting. . . . They see how a manufacturing plant is going to introduce new equipment. . . .
They understand the quality control systems that are
over the business. . . . They see the whole company
and there's an advantage to that.

Mr. Dinkins's statement makes sense if we are looking for
tactical, finite-minded leadership. But not if what we need is
a CVO. A CVO is not an operations or a finance job. Whereas
CVOs focus on up and out, CFOs and COOs focus on down
and in. One requires eyes on the infinite horizon, the other
requires eyes on the business plan. One envisions the very distant, abstract future. The other sees the steps to take in the
tangible near term.

This is one of the reasons the best organizations are often
run in tandem. The combination of the keeper of the vision
(CVO) and the operator (the CFO or COO). It is a partnership of complementary skill sets. We are more likely to get
these partnerships if we adjust the formal hierarchies in our
companies to promote the right mindset to fit the purpose of
the job. This means that we need to stop seeing the CEO as
number one and the CFO or COO as number two and start
thinking of them as vital partners in a common cause. One
does not know how to do the other's job better than they do
(which is why they need each other). Remember, Steve Ballmer, John Sculley and Kevin Rollins all thrived when they
were working alongside their more infinite-minded partners.

Though the CVO is more often in the spotlight, and
though the CVO is often given more of the praise, publicly at
least, both players must have the strength of ego to know it is
a trusted partnership. The CVO knows they cannot advance
their vision alone and need someone like Michael Dinkins

described by their side. The COO or CFO knows that their skills can work to vastly greater scale and meaning if they are applied to help advance an infinite Just Cause; something bigger than themselves or the company. Such a model has precedence. In the military there are officers and enlisted ranks who work alongside each other. To rise in the enlisted ranks is a different trajectory than a rise in the officer ranks. They are entirely different career paths. There is no conflict of interest when they work together because the most senior enlisted leader on a base cannot aspire to take the job of the most senior officer, and vice versa. When these partnerships work, the CVO and the COO or the CFO spend more time thanking and celebrating each other than competing for attention.

An uncomfortable truth for many CFOs or COOs is that they have already reached the top level of their skill set. They are already the most senior, most skilled finance or operations people in the organization, which is a great thing. Without them, the CVO would not be able to advance the vision. But that doesn't mean that they are equipped to be at the forefront, leading that vision. For many, once they get the "top job" they are more likely to continue doing what they know and do well—thinking about how big they want their companies to be and what kinds of margins, EBITDA, EPS or market share they aim to achieve (finite pursuits)—than they are to embrace the new responsibility of imagining what the future could look like and how the company might advance a Just Cause (an infinite pursuit).

It's like a salesperson who is promoted to sales manager. They might have excelled at making sales, but they are no longer responsible for selling; they are now responsible for taking care of the people who do the selling. If they fail to shift gears, adjust their mindset and learn a new set of skills for their new responsibility, problems will ensue. Any CFO, COO or other executive can absolutely succeed as CVO if they also learn to adapt to their new role and new responsibilities and embrace an infinite mindset. If they fail to do so, they will likely default to the skills that got them their previous job, which increases the probability that they will steer the company down a very finite path.

Whether or not he was qualified to be CVO of Walmart, Duke failed to adjust for the role he was given—he failed to champion Sam Walton's vision into the next century. In contrast, Duke's successor, Doug McMillon, could prove to be the CVO that Walmart needs. When his new position was announced in 2013, McMillon said in a press release, "The opportunity to lead Walmart is a great privilege. Our company has a rich history of delivering value to customers across the globe and, as their needs grow and change, we will be there to serve them. Our management team is talented and experienced, and our strategy gives me confidence that our future is bright. By keeping our promise to customers, we will drive shareholder value, create opportunity for our associates and grow our business." McMillon presented his priorities in literally the exact opposite order that Mike Duke had when he stepped up to lead the company five years earlier. McMillon put Sam Walton's vision first. It is exciting to see how he is reequipping Walmart to once again play in the Infinite Game.

Chapter 5

THE RESPONSIBILITY OF BUSINESS (REVISED)

Business today is subject to a dizzying rate of change. And all that change seems to be taking its toll. The time it takes before a company is forced out of the game is getting shorter and shorter. The average life of a company in the 1950s, if you recall, was just over 60 years. Today it is less than 20 years. According to a 2017 study by Credit Suisse, disruptive technology is the reason for the steep decline in company life span. However, disruptive technologies are not a new phenomenon. The credit card, the microwave oven, Bubble Wrap, Velcro, transistor radio, television, computer hard disks, solar cells, optic fiber, plastic and the microchip were all introduced in the 1950s. Save for Velcro and Bubble Wrap (which are disruptive in a completely different way), that's a pretty good list of disruptive technologies. "Disruption" is likely not the cause of the challenge, it's a symptom of a more insidious root cause. It is not technology that explains failure; it is less about technology, per se, and more about the leaders' failure to envision the future of their business as the world changes around them. It is the result of shortsightedness. And shortsightedness is an inherent condition of leaders who play with a finite mindset. In fact, the rise of this kind of shortsightedness over the past 50 years can be traced back to the philosophies of a single person.

In a watershed article from 1970, Milton Friedman, the Nobel Prize–winning economist, who is considered one of the great theorists of today's form of capitalism, laid out the foundation for the theory of shareholder primacy that is at the heart of so much finite-minded business practice today. "In a free-enterprise, private-property system," he wrote, "a corporate executive is an employee of the owners of the business. He has direct responsibility to his employers. That responsibility is to conduct the business in accordance with their desires, which generally will be to make as much money as possible while conforming to the basic rules of the society, both those embodied in law and those embodied in ethical custom." Indeed, Friedman insisted that "there is one and only one social responsibility of business, to use its resources and engage in activities designed to increase its profits so long as it stays within the rules of the game." In other words, according to Friedman, the sole purpose of business is to make money and that money belongs to shareholders. These ideas are now firmly ingrained in the zeitgeist. Today it is so generally accepted that the "owner" of a company sits at the top of the benefit food chain and that business exists solely to create wealth, that we often assume that this was always the way that the game of business was played and is the only way it can be played. Except it wasn't . . . and it isn't.

Friedman seemed to have a very one-dimensional view of business. And as anyone who has ever led, worked for or bought from a business knows, business is dynamic and complicated. Which means, it is possible that, for the past 40+ years, we have been building companies with a definition of business that is actually bad for business and undermines the very system of capitalism it proclaims to embrace.

Capitalism Before Friedman

For a more infinite-minded alternative to Friedman's definition of the responsibility of business, we can go back to Adam Smith. The eighteenth-century Scottish philosopher and economist is widely accepted as the father of economics and modern capitalism. "Consumption," he wrote in *The Wealth of Nations*, "is the sole end and purpose of all production and the interest of the producer ought to be attended to, only so far as it may be necessary for promoting that of the consumer." He went on to explain, "The maxim is so perfectly self-evident, that it would be absurd to attempt to prove it." Put simply, the company's interests should always be secondary to the interest of the consumer (ironically, a point Smith believed so "self-evident," he felt it was absurd to try to prove it, and yet here I am writing a whole book about it).

Smith, however, was not blind to our finite predilections. He recognized that "in the mercantile system the interest of the consumer is almost constantly sacrificed to that of the producer; and it seems to consider production, and not consumption, as the ultimate end and object of all industry and commerce." In a nutshell, Smith accepted that it was human nature for people to act to advance their own interests. He called our propensity for self-interest the "invisible hand." He went on to theorize that because the invisible hand was a universal truth (because of our selfish motivations we all want to build strong companies), it ultimately benefits the consumer. "It is not from the benevolence of the butcher, the brewer, or the baker that we can expect our dinner, but from their regard to their own interest," he explained. The butcher has a selfish desire to offer the best cuts of meat without regard for the brewer or the baker. And the brewer wants to make the best beer, regardless of what meat or bread is available on the market. And the baker wants to make the tastiest loaves without any consideration for what we may put on our sandwiches. The result, Smith believed, is that we, the consumers, get the best of everything . . . at least we do if the system is balanced. However, Smith did not consider a time in which the selfishness of outside investors and an analyst community would put that system completely out of balance. He did not anticipate that an entire group of self-interested outsiders would exert massive pressure on the baker to cut costs and use cheaper ingredients in order to maximize the investors' gains.

If history or 18th-century brogue-tongued philosophers are not your jam, we need simply look at how capitalism changed after the idea of shareholder supremacy took over—which only happened in the final decades of the twentieth century. Prior to the introduction of the shareholder primacy theory, the way business operated in the United States looked quite different. "By the middle of the 20th century," said Cornell corporate law professor Lynn Stout in the documentary series *Explained*, "the American public corporation was proving itself one of the most effective and powerful and beneficial organizations in the world." Companies of that era allowed for average Americans, not just the wealthiest, to share in the investment opportunities and enjoy good returns. Most important, "executives and directors viewed themselves as stewards or trustees of great public institutions that were supposed to serve not just the shareholders, but also bondholders, suppliers, employees and the community." It was only after Friedman's 1970 article that executives and directors started to see themselves as responsible to their "owners," the shareholders, and not stewards of something bigger. The more that idea took hold in the 1980s and '90s, the more incentive structures inside public companies and banks themselves became excessively focused on shorter-and-shorter-term gains to the benefit of fewer and fewer people. It's during this time that the annual round of mass layoffs to meet arbitrary projections became an accepted and common strategy for the first time. Prior to the 1980s, such a practice simply didn't exist. It was common for people to work a practical lifetime for one company. The company took care of them and they took care of the company. Trust, pride and loyalty flowed in both directions. And at the end of their careers these long-time employees would get their proverbial gold watch. I don't think getting a gold watch is even a thing anymore. These days, we either leave or are asked to leave long before we would ever earn one.

Capitalism Abuse

The finite-minded form of capitalism that exists today bears little resemblance to the more infinite-minded form that inspired America's founders (Thomas Jefferson owned all three volumes of Smith's *Wealth of Nations*) and served as the bedrock for the growth of the American nation. Capitalism today is, in name only, the capitalism that Adam Smith envisioned over 200 years ago. And it looks nothing like the capitalism practiced by companies like Ford, Kodak and Sears in the late 19th and early 20th centuries, before they too fell prey to finite thinking and lost their way. What many leaders in business practice these days is more of an abuse of capitalism, or "capitalism abuse." Like in the case of alcohol abuse, "abuse" is defined as improper use of something. To use something for a reason other than that for which it was intended. And if capitalism was intended to benefit the consumer and the leaders of companies were to be the stewards of something greater than themselves, they are not using it that way today.

Some may say my view—that the purpose of a company is not just to make money but to pursue a Just Cause—is naïve and anticapitalist. First, I would urge us all to beware the messenger. My assumption is that those who most fiercely defend Friedman's views on business, and many of the current and accepted business practices he inspired, are the ones who benefit most from them. But business was never just about making money. As Henry Ford said, "A business that makes nothing but money is a poor kind of business." Companies exist to advance something—technology, quality of life or anything else with the potential to ease or enhance our lives in some way, shape or form. That people are willing to pay money for whatever a company has to offer is simply proof that they perceive or derive some value from those things. Which means the more value a company offers, the more money and the more fuel they will have for further advancements. Capitalism is about more than prosperity (measured in features and benefits, dollars and cents); it's also about progress (measured in quality of life, technological advancements and the ability of the human race to live and work together in peace).

The constant abuse since the late 1970s has left us with a form of capitalism that is now, in fact, broken. It is a kind of bastardized capitalism that is organized to advance the interests of a few people who abuse the system for personal gain, which has done little to advance the true benefits of capitalism as a philosophy (as evidenced by anticapitalist and protectionist movements around the globe). Indeed, the entire philosophy of shareholder primacy and Friedman's definition of the purpose of business was promoted by investors themselves as a way to incentivize executives to prioritize and protect their finite interests above all else.

It is due in large part to Milton Friedman's ideas, for example, that corporations started tying executive pay to short-term share price performance rather than the long-term health of the company. And those who embraced Friedman's views rewarded themselves handsomely. The Economic Policy Institute reported that in 1978, the average CEO made approximately 30 times the average worker's salary. By 2016, the average had increased over 800 percent to 271 times the average worker's pay. Where the average CEO has seen a nearly 950 percent increase in their earnings, the American worker, meanwhile, has seen just over 11 percent in theirs. According to the same report, average CEO pay has increased at a rate 70 percent faster than the stock market!

It doesn't take an MBA to understand why. As Dr. Stout explains in her book, *The Shareholder Value Myth*, "If 80 percent of the CEO's pay is based on what the share price is going to do next year, he or she is going to do their best to make sure that share price goes up, even if the consequences might be harmful to employees, to customers, to society, to the environment or even to the corporation itself in the long-term." When we tie pay packages directly to stock price, it promotes practices like closing factories, keeping wages down, implementing extreme cost cutting and conducting annual rounds of layoffs—tactics that might boost the stock price in the near term, but often do damage to an organization's ability to survive and thrive in the Infinite Game. Buy-backs are another often legitimate practice that has been abused by public company executives seeking to prop up their share price. By buying back its own shares, based on the laws of supply and demand, they temporarily increase demand for their stock, which temporarily drives up the price (which temporarily makes the executives look good).

Though many of the practices used to drive up stock prices in the short term sound ethically dubious, if we look back to Friedman's definition of the responsibility of business, we find that he leaves the door wide open for such behavior, even encourages it. Remember, his only guidance for the responsibility companies must obey is to act within the bounds of the law and "ethical custom." I, as one observer, am struck by that awkward phrase, "ethical custom." Why not just say "ethics"? Does ethical custom mean that if we do something frequently enough it becomes normalized and is thus no longer unethical? If so many companies use regular rounds of mass layoffs, using people's livelihoods, to meet arbitrary projections, does that strategy then cease to be unethical? If everyone is doing it, it must be okay.

As a point of fact, laws and "ethical customs" usually come about in response to abuses, not by predicting them. In other words, they always lag behind. Based on the common interpretation of Friedman's definition, it's almost a requirement for companies to exploit those gaps to maximize profit until future laws and ethical customs tell them they can't. Based on Friedman, it is their responsibility to do so!

Technology companies, like Facebook, Twitter and Google, certainly look like they are more comfortable asking for forgiveness as they run afoul of ethical customs, as opposed to leading with a fundamental view of how they safeguard one of their most important assets: our private data. Based on Friedman's standards, they are doing exactly what they should do.

If we are using a flawed definition of business to build our companies today, then we are likely also promoting people and forming leadership teams best qualified to play by the finite rules that Friedman espoused—leadership teams that are probably the least equipped to navigate the ethical requirements necessary to avoid exploiting the system for self-gain. Built with the wrong goal in mind, these teams are more likely to make decisions that do long-term damage to the very organizations, people and communities they are supposed to be leading and protecting. As King Louis XV of France said in 1757, *"Après moi le dèluge."* "After me comes the flood." In other words, the disaster that will follow after I'm gone will be your problem, not mine. A sentiment that seems to be shared by too many finite leaders today.

The Pressure to Play with a Finite Mindset

It's a big open secret among the vast majority of public-company executives that the theory of shareholder primacy and the pressure Wall Street exerts on them are actually bad for business. The great folly is that despite this knowledge and their private grumblings and misgivings, they continue to defend the principle and yield to the pressure.

I am not going to waste precious ink making a drawn-out argument about the long-term impact of what happened to our country and global economies when executives bowed to those pressures. It is enough to call attention to the man-made recession of 2008, the increasing stress and insecurity too many of us feel at work and a gnawing feeling that too many of our leaders care more about themselves than they do about us. This is the great irony. The defenders of finite-minded capitalism act in a way that actually imperils the survival of the very companies from which they aim to profit. It's as if they have decided that the best strategy to get the most cherries is to chop down the tree.

Thanks in large part to the loosening of regulations that were originally introduced to prevent banks from wielding the kind of influence and speculative tendencies that caused the Great Depression of 1929 to happen, investment banks once again wield massive amounts of power and influence. The result is obvious—Wall Street forces companies to do things they shouldn't do and discourages them from doing things they should.

Entrepreneurs are not immune from the pressure either. In their case, there is often intense pressure to demonstrate constant, high-speed growth. To achieve that goal, or when growth slows, they turn to venture capital or private equity firms to raise money. Which sounds good in theory. Except there is a flaw in the business model of private equity that can wreak havoc with any company keen to stay in the game. For private equity and venture capital firms to make money, they have to sell. And it's often about three to five years after they make their initial investment. A private equity firm or venture capitalist can use all the flowery, infinite game, Cause-focused language they want. And they may believe it.

Up until the point they have to sell. And then all of a sudden many will care a lot less about the Just Cause and all the other stakeholders. The pressure investors can exert on the company to do things in the name of finite objectives can be and often is devastating to the long-term prospects of the company. Long is the list of purpose-driven executives who say that *their* investors are different, that they *do* care about the company's Cause . . . until it's time to sell. (The ones I talked to asked that I not mention the names of their companies for fear of upsetting their investors.)

There is no such thing as constant growth, nor is there any rule that says high-speed growth is necessarily a great strategy when building a company to last. Where a finite-minded leader sees fast growth as the goal, an infinite-minded leader views growth as an adjustable variable. Sometimes it is important to strategically slow the rate of growth to help ensure the security of the long-term or simply to make sure the organization is properly equipped to withstand the additional pressures that come with high-speed growth. A fast-growing retail operation, for example, may choose to slow the store expansion schedule in order to put more resources into training and development of staff and store managers. Opening stores is not what makes a company successful; having those stores operate well is. It's in a company's interest to get things done right now rather than wait to deal with the problems high-speed growth can cause later. The art of good leadership is the ability to look beyond the growth plan and the willingness to act prudently when something is not ready or not right, even if it means slowing things down.

From the 1950s to the '70s, the concept of "forecasting" was considered critical across multiple institutions. Teams of "futurists" were brought in to examine technological, political and cultural trends in order to predict their future impact and prepare for it. (Such a practice may have helped Garmin proactively adapt to advancements in mobile phone technology instead of being forced to react to it.) Even the United States federal government was in on it. In 1972, Congress established the Office of Technology Assessment specifically to examine the long-term impact of proposed legislation. "They're beginning to realize that legislation will remain on the books for 20 or 50 years before it's reviewed," said Edward Cornish, president of the World Future Society, "and they want to be sure that what they do now won't have an adverse impact years from today." However, the discipline fell out of favor during the 1980s, with some in government thinking it a waste of money to try to "predict the future." The office was officially closed in 1995. Though today futurists still exist in the business world, they are usually tasked with helping a company predict trends that can be marketed to rather than assessing future impact of current choices.

Finite-focused leaders are often loath to sacrifice near-term gains, even if it's the right thing to do for the future, because near-term gains are the ones that are most visible to the market. And the pressure this mindset exerts on others in the company to focus on the near-term often comes at the detriment of the quality of the services or the products we buy. That is the exact opposite of what Adam Smith was talking about. If the investor community followed Smith's philosophies, they would be doing whatever they could to help the companies in which they invested make the best possible product, offer the best possible service and build the strongest possible company. It's what's good for the customer and the wealth of nations. And if shareholders really were the owners of the companies in which they invested, that is indeed how they would act. But in reality, they don't act like owners at all. They act more like renters.

Consider how differently we drive a car we own versus one we rent, and all of a sudden it will become clear why shareholders seem more focused on getting to where they want to go with little regard to the vehicle that's taking them there. Turn on CNBC on any given day and we see discussions dominated by talk of trading strategies and near-term market moves. These are shows about trading, not about owning. They are giving people advice on how to buy and flip a house, not how to find a home to raise a family. If short-term-focused investors treat the companies in which they invest like rental cars, i.e., not theirs, then why must the leaders of the companies treat those investors like owners? The fact is, public companies are different from private companies and do not need to conform to the same traditional definition of ownership. If our goal is to build companies that can keep playing for lifetimes to come, then we must stop automatically thinking of shareholders as owners, and executives must stop thinking that they work solely for them. A healthier way for all shareholders to view themselves is as contributors, be they near-term or long-term focused.

Whereas employees contribute time and energy, investors contribute capital (money). Both forms of contribution are valuable and necessary to help a company succeed, so both parties should be fairly rewarded for their contributions. Logically, for a company to get bigger, stronger or better at what they do, executives must ensure that the benefit provided by investors' money or employees' hard work should, as Adam Smith pointed out, go first to those who buy from the company. When that happens, it is easier for the company to sell more, charge more, build a more loyal customer base and make more money for the company and its investors alike. Or am I missing something here? In addition, executives need to go back to seeing themselves as stewards of great institutions that exist to serve all the stakeholders. The impact of which serves the wants, needs and desires of all those involved in a company's success, not just a few.

The fact is, we all want to feel like our work and our lives have meaning. It's part of what it means to be human. We all want to feel a part of something bigger than ourselves. I have to believe this contributes to the reason so many companies say they primarily serve their people and their customers when they are in fact primarily serving their executive ranks and their shareholders. For many of us, even if we don't have the words, the modern form of capitalism we have just feels like something doesn't align with our values. Indeed, if we all truly embraced Friedman's definition of business, then companies would have visions and missions that were solely about maximizing profit and we'd all be fine with it. But they don't. If the true purpose of business was only to make money, there would be no need for so many companies to pretend to be cause or purpose driven. Saying a business exists for something bigger and actually building a business to do it are not the same thing. And only one of those strategies has any value in the Infinite Game.

The Drums of Change Are Beating

In 2018, Larry Fink, the founder, chairman and CEO of BlackRock, Inc., caused a bit of a stir in the financial industry when he wrote an open letter to CEOs titled "A Sense of Purpose." In the letter he urged leaders to build their companies with more idealistic goals than near-term financial gains.

"Without a sense of purpose," he explained, "no company, either public or private, can achieve its full potential. It will ultimately lose the license to operate from key stakeholders. It will succumb to short-term pressures to distribute earnings, and, in the process, sacrifice investments in employee development, innovation, and capital expenditures that are necessary for long-term growth." BlackRock, incidentally, is the largest money management firm in the world, with over $6 trillion under management. Though the call for companies to embrace a sense of purpose is not new, when someone of Larry Fink's position in the financial world embraces the concept so publicly, it moves the conversation from articles, books and water coolers to inside palace walls.

The stock market works at its best when it works as it was intended, to allow for the average person to share in the wealth of the nation. However, Americans have become disillusioned with the form of capitalism to which they are subjected today and the way the stock market is used as a tool in a finite game. The share of Americans invested in the stock market is at its lowest point in 20 years. The largest exodus has come from the middle class. People don't mind if an enterprising few make a lot of money. Their exodus is a reaction to the imbalance and a lack of trust in the system . . . and leaders should take notice.

The irony is that everyone who works with or for the public markets understands that when the system becomes too unbalanced, there will always be a correction. That correction is often sudden and violent. Our current system of capitalism is so unbalanced, and those on the inside are well advised to make the necessary corrections themselves, for a failure to do so increases the chances of correction being forced upon them. For if the palace refuses to change from within, it increases the chances that the people will try to knock the whole thing down. Be they against government incompetence, corruption or lopsided economic models, this is what populist uprisings are so often about. Remember the American Revolution itself would have been avoided if Great Britain simply relaxed the economic restrictions it placed on the colonies, gave them greater representation in government and allowed them to share in more of the wealth they helped produce. That's it. Where there is unbalance, there is unrest.

It's a big deal to disrupt a system. Revolutions are fraught with risk. They are sudden. They are violent. And there is almost always a counterrevolution (and when I talk about revolution, I am not only referring to armed insurgencies, I include all kinds of upending to the status quo). The American colonists chose to revolt only after years of appealing for change. Begging for it. They were only partially drawn to revolution for ideological reasons. They were pushed to it because they saw their lives and their economic well-being suffering or restricted as a result of a gross imbalance of power and wealth. The vision of an alternative future came later.

Whether it was in ancient Rome, where the leaders refused to offer citizenship to the allies who suffered to defend Rome, or the American colonists who were refused representation even though their hard work helped fuel the British economy, it is upon the backs of ordinary people that wealth and power are produced. In our modern day and age, it is the employee who bears the most cost for the money companies and their leaders make. They are the ones who must worry every time the company misses its arbitrary projections whether they will be sent home without the means to provide for themselves or their families. It is the employee who comes to work and feels that the company and its leaders do not care about them as human beings (note: offering free food and fancy offices is not the thing that makes people feel cared for). People want to be treated fairly and share in the wealth they helped produce in payment for the cost they bear to grow their companies. I am not demanding it—they are!

The data shows that the current system benefits the top 1 percent of the population disproportionately more than anyone else. In response to that imbalance, a small group of protesters set up camp in Zuccotti Park in New York City in September 2011. They posted signs that said simply, "We are the 99 percent." Leaderless and unfocused, the occupation of parks around the world fizzled but the movement lives on. The spotlight on the fact that the system was rigged for the few at the expense of the masses has not dimmed. If anything, it has grown brighter. Five years since the start of the Occupy movement we heard the populist message rise to the level of a presidential election from Bernie Sanders on the left and Donald Trump on the right. Both candidates fanned the flames about inequality and unfairness of "the system."

The call to abandon Milton Friedman's style of business, like any challenge to any status quo, can come from the people or from the leaders. From outside or from inside. Take heed of the red flags all around us. The rise of a populist voice in America and around the world is growing. And all those in a seat of power—be they in business or in politics—are in a position to effect change. But make no mistake, change is coming. Because that's how the Infinite Game works. This finite system we have now will run itself dry of will and resources eventually. It always does. It always does. Though some may enrich themselves with money or power for now, the system cannot survive under its own weight. If history and almost every stock market crash is any indicator, imbalance is a bitch.

The winds of change are blowing. It has become more socially acceptable to question some of the accepted tenets of Friedman's capitalism. And there continues to be a growing discomfort with such devotion to his definition of the responsibility of business. Organizations like Conscious Capitalism, B Corp, the B Team and others are actively promoting ideas like the stakeholder model or triple bottom line, to challenge Friedman's ideas. And the business heroes of the high flying 1980s and '90s, like Jack Welch, are losing their luster and appeal. It is now self-evident that we need a new definition of the responsibility of business that better aligns with the idea that business is an infinite game. A definition that understands that money is a result and not a purpose. A definition that gives employees and the people who lead them the feeling that their work has value beyond the money they make for themselves, their companies or their shareholders.

Friedman proposed that a business has a single responsibility—profit; a very finite-minded view of business. We need to replace Friedman's definition with one that goes beyond profit and considers the dynamism and additional facets that make business work. In order to increase the infinite value to our nation, our economy and all the companies that play in the game, the definition of the responsibility of business must:

1. Advance a purpose: Offer people a sense of belonging and a feeling that their lives and their work have value beyond the physical work.
2. Protect people: Operate our companies in a way that protects the people who work for us, the people who buy from us and the environments in which we live and work.

3. Generate profit: Money is fuel for a business to remain viable so that it may continue to advance the first two priorities.

Simply put:

The responsibility of business is to use its will and resources to advance a cause greater than itself, protect the people and places in which it operates and generate more resources so that it can continue doing all those things for as long as possible. An organization can do whatever it likes to build its business so long as it is responsible for the consequences of its actions.

The three pillars—to advance a purpose, protect people and generate a profit—seem to be essential in the Infinite Game. America's founders inspired a nation to come together to advance Life, Liberty and the pursuit of Happiness. These unalienable rights of physical safety, a cause or ideology to be a part of and the opportunity to provide for ourselves inspired a nation and set the United States on its infinite journey. Nearly 150 years later, on December 30, 1922, the Declaration of the Formation of the Soviet Union was ratified. It stated that the new nation of the USSR was founded on the three promises or rights: "All these circumstances imperatively demand the unification of the Soviet republics into one union state, capable of ensuring both external security and internal economic prosperity, and the freedom of the national development of peoples." In other words, a nation committed to protect its people, offer an opportunity of economic gain and advance the ideology of communism. A similar trifecta showed up again during the Vietnam War when General

Giap rallied the North Vietnamese to join the People's War with the promise of physical safety, economic advancement and the opportunity to advance an ideology. A People's War is "simultaneously military, economic and political," said Giap in an interview years after the war.

A nation state must protect its citizens, to ensure that we live free from fear. To do that, it must maintain armed forces to defend against foreign threats, establish justice and insure domestic tranquillity. Likewise, inside an organization, a company must provide for the protection of its people by building a culture in which employees feel psychologically safe and feel like their employer cares about them as human beings. We want to know that the company is invested in our growth as much as it is its own. No one should have to come to work in fear of the annual round of layoffs simply because the company missed an arbitrary projection. A company can provide for the safety and protection of those outside its walls by considering how the manufacturing of its products and the ingredients they choose impact the communities in which those products are made or sold.

For nations, our sense of belonging and ideologies that we would sacrifice to advance often come in the form of -isms, like capitalism, socialism and so on. In business, they come in the form of a Just Cause. In both the place we choose to live and the place we choose to make a living, we should feel like we are working to advance something bigger than ourselves.

Among nations, profit matters. Economic prosperity is the ability for the nation to remain solvent. To maintain a strong economy that is well resourced to thrive in good times and survive in lean times. For businesses, it is the same. And both in nations and in companies, everyone wants the opportunity to work hard and earn an income so that we may provide for ourselves and our families.

The goals of a nation founded with an infinite mindset are also the people's goals. A nation exists to serve and include ordinary people as it strives forward. This is what makes us feel emotionally connected to our country, why we feel patriotic. Translated into business terms, it means that a company's goals must also align with people's goals, not simply the goals of shareholders. If we want our work to benefit ourselves, our colleagues, our customers, our communities and the world, then it is right for us to work at companies whose values and goals align with our own. And if they don't, we can demand that they do. Anyone who offers their blood, sweat and tears to advance a company's goals is entitled to feel valued for their contributions and share in the fruits of their labor.

Where Friedman believed the results of our hard work should be for the primary benefit of an elite ruling class (the owner), the more infinite-minded leader would ensure that, so long as there are shared goals, all who contribute will benefit across all three pillars. We are all entitled to feel psychologically protected at work, be fairly compensated for our effort and contribute to something bigger than ourselves. These are our unalienable rights. Business, like any infinite pursuit, is a more powerful force when it is empowered for the people, by the people. Disruption is not going away anytime soon, that's not going to change. How leaders respond to it, however, can. Where Friedman's finite definition of the responsibility focuses on maximizing resources, a revised infinite definition also considers the will of the people.

Chapter 6

WILL AND RESOURCES

The Four Seasons in Las Vegas is a wonderful hotel. The reason it's a wonderful hotel is not because of the fancy beds. Any hotel can buy fancy beds. The reason the Four Seasons is a wonderful hotel is because of the people who work there. If you find yourself walking through the halls and an employee says hello, for example, you get the distinct feeling that they wanted to say hello, not that they were told to say hello. Human beings are highly attuned social animals; we can tell the difference.

There happens to be a coffee bar in the lobby of the hotel. One afternoon while on a business trip in Las Vegas, I went to buy myself a cup of coffee. The barista working that day was a young man named Noah. Noah was funny and engaging. It was because of Noah that I enjoyed buying that cup of coffee more than I generally enjoy buying a cup of coffee. After standing and chatting for a while, I finally asked him, "Do you like your job?" Without skipping a beat Noah immediately replied, "I love my job!"

Now, for someone in my line of business, that's a significant response. He didn't say, "I like my job," he said, "I love my job." That's a big difference. "Like" is rational. We like the people we work with. We like the challenge. We like the work. But "love," love is emotional. Love is something harder to quantify. It's like asking someone "Do you love your spouse," and they respond, "I like my spouse a lot." It's a very different answer. You get my point, love is a higher standard. So when Noah said, "I *love* my job," I perked up. From that one response, I knew Noah felt an emotional connection to the Four Seasons that was bigger than the money he made and the job he performs.

Immediately, I asked Noah a follow-up question. "Tell me specifically what the Four Seasons is doing that you would say to me that you *love* your job." Again without skipping a beat, Noah replied, "Throughout the day, managers will walk past me and ask me how I'm doing, ask me if there is anything I need, anything they can do to help. Not just my manager . . . *any* manager. I also work for [another hotel]," he continued. He went on to explain that at his other job the managers walk past and try to catch people doing things wrong. At the other hotel, Noah lamented, "I keep my head below the radar. I just want to get through the day and get my paycheck. Only at the Four Seasons," Noah said, "do I feel I can be myself."

Noah gives his best when he's at the Four Seasons. Which is what every leader wants from their people. So it makes sense why so many leaders, even some of the best-intentioned ones, often ask, "How do I get the most out of my people?" This is a flawed question, however. It's not a question about how to help our people grow stronger, it's about extracting more output from them. People are not like wet towels to be wrung out. They are not objects from which we can squeeze every last drop of performance. The answers to such a question might yield more output for a time, but it often comes at a cost of our people and to the culture in the longer term. Such an approach will never generate the feelings of love and commitment that Noah has for the Four Seasons. A better question to ask is, "How do I create an environment in which my people can work to their natural best?"

Too often, when performance lags, the first thing we do is blame the people. But in Noah's case, he is the same person in both his jobs. The only difference is the leadership environment in which he is asked to work. Had I met Noah at the other hotel, where his output was prioritized over how supported he felt, my experience with him would have been totally different. The odds are high I would neither be writing about him nor singing the praises of the other hotel. It's not the people doing the job, it's the people who lead the people doing the job who can make the greater difference.

Noah's managers at the Four Seasons understand that their job is to set an environment for Noah in which he can naturally thrive. Leaders will work to create these environments when we train them how to prioritize their people over the results. And this is the true definition of what it means to lead. There is absolutely zero cost for a manager to take time to walk the halls and ask their people how they are doing . . . and actually care about the answers. Because the leadership at the Four Seasons prioritizes the will of their people before the resources they can produce, the people who work there *want* to give their jobs their all and the guests of the Four Seasons can feel it.

Will Before Resources

In any game, there are always two currencies required to play—will and resources. Resources are tangible and easily measured. When we talk about resources, we're usually talking about money. And depending on an organization's preferences or the standards of the day, those resources can be counted in multiple ways—revenues, profit, EBITDA, EPS, cash flow, venture capital, private equity, stock price and so on. Resources generally come from outside sources, like customers or investors, and represent the sum of all the financial metrics that contribute to the health of the organization.

Will, in contrast, is intangible and harder to measure. When we talk about will, we're talking about the feelings people have when they come to work. Will encompasses morale, motivation, inspiration, commitment, desire to engage, desire to offer discretionary effort and so on. Will generally comes from inside sources like the quality of leadership and the clarity and strength of the Just Cause. Will represents the sum of all the human elements that contribute to the health of the organization.

All leaders, whether operating with a finite or infinite mindset, know resources are essential. And both finite- and infinite-minded leaders agree that will is also essential. I have yet to meet any CEO who thinks their people are unimportant. The problem is, will and resources can never be equally prioritized. There are always circumstances in which one is pitted against the other, times in which a leader must choose which one they are willing to sacrifice. The question is, which one will they choose? Every leader has a bias.

Most of us have sat in a meeting and listened to a leader present their priorities . . . and it often looks something like this: 1. Growth. 2. Our customers. 3. Our people. Though that leader will insist that they do care about their people ("our people" is one of their priorities), the order in which they appear on the list matters. In this case, there are at least two things that are considered more important than the people, and one of them is resources. How a leader lists their priorities reveals their bias. And their bias will influence the choices they make.

The finite-minded leader tends to show a bias for the score. As a result, they often opt for choices that demonstrate results in a short time frame, even if doing so, "regrettably," comes at a cost to the people. These are leaders who, during hard times, for instance, will turn first to layoffs and extreme cost cutting measures rather than explore alternatives that may not demonstrate the same immediate results, even if they have longer-term benefits. If a leader has a bias for resources, it is much easier for them to calculate the immediate savings of reducing 10 percent of their workforce next week than it is to choose an option in which the savings take longer to hit the balance sheet.

Infinite-minded leaders, in contrast, work hard to look beyond the financial pressures of the current day and put people before profit as often as possible. In hard times, they are less likely to look at their people as just another expense to be cut and more willing to explore other ways to save money, even if the results may take longer to realize. The infinite-minded leader may opt for furloughs instead of lay-offs to help manage the resources; for example, requiring every employee to take two or three weeks of unpaid time off. Though people may be asked to sacrifice some money, everyone keeps their job. When a group shares in the suffering, it actually brings a team together. It is the same reason people come together after a natural disaster. However, when some are forced to bear an unbalanced amount of the burden, it can rip a culture apart. Thinking beyond the hard times, an infinite-minded leader is okay to wait the quarter or the year or more for the savings to accumulate if it means safeguarding the will of the people. They understand that the will of their people is *the* thing that drives discretionary effort, as well as problem solving, imagination and teamwork—all things essential for surviving and thriving in the future. The value of strong will over resources simply cannot be underestimated. Indeed, it was the will of the North Vietnamese people that was central to General Giap's strategy to push the superior-resourced American forces out of Vietnam.

Still, when those with a bias for the resources hear folks like me talk about the need to put people before profit, the hair on the back of their necks stands up. What they hear is that I think the money is not important. False. What they hear is that I don't think they care about their people. Also false. It's not an either-or choice. The bias doesn't even need to be extreme. Danny Meyer, the famed restaurateur and founder of Shake Shack, shared his bias when he said his business is 49 percent technical and 51 percent emotional (the restaurateur's take on will and resources). Even a small bias for will before resources is more likely to create a stronger culture in which will and resources will both be in ample supply for the long game.

The Cost of Will

Too many leaders "see people as a cost," says former CEO of Burberry and former senior vice president of retail for Apple Angela Ahrendts. Especially in retail, which suffers from such high turnover rates, the common logic is, "Why invest in people who aren't gonna stick around?" This is a one-dimensional and finite view of the way business works. Focusing on the money they can save by not investing in their people, too many finite-minded leaders overlook the additional costs they actually incur when they don't. Hiring new people to fill the empty slots costs money. Losing experienced staff and waiting for people to get trained and adjust to a new culture all affect productivity. Add in the low morale in high-turnover jobs, and it makes one curious whether the money saved was actually worth it. Ahrendts was curious too. So she ran the numbers. And what she discovered surprised even her. The actual incremental cost of Apple taking care of their people was: zero.

Apple gives all full-time retail employees the same benefits as full-time employees who work at corporate, including full medical and dental coverage and $2,500 in education reimbursement should they wish to take classes outside work. Apple was one of the first companies to offer new hires a $15-per-hour minimum wage and gives full-time retail employees the same option to buy stock in the company as any other corporate employee. All these additional costs are offset by the money the company saves from lower recruiting and training costs, which most firms that overuse layoffs are forced to pay to refill positions at later dates (costs that are often not included when executives report how much money they saved with a round of layoffs). And unlike many large retailers who have to maintain a huge staff of recruiters to work continually to replace the people who leave, Apple only needs a very lean recruiting staff for their retail operations. Of course, some would argue that Apple makes a lot more money per employee compared to most retail operations and so they can afford to pay higher wages. However, Costco, which pays their cashiers an average of $15.09 (in addition to offering a 401(k) and health insurance), has found that they make up for the additional cost because of reduced turnover and higher productivity. Plus, customers tend to enjoy better service when employees feel looked after, which likely translates into higher average sales.

If the actual costs are net neutral, then the difference in how we treat people is simply a matter of mindset. And it is because of that alternative mindset that Apple and Costco enjoy average retention rates around 90 percent, when the average in the rest of retail is 20 to 30 percent. Where finite-minded organizations view people as a cost to be managed, infinite-minded organizations prefer to see employees as human beings whose value cannot be calculated as if they were a piece of machinery. Investing in human beings goes beyond paying them well and offering them a great place to work. It also means treating them like human beings. Understanding that they, like all people, have ambitions and fears, ideas and opinions and ultimately want to feel like they matter. It may feel like a risk to many a finite-minded leader. To shell out all that extra money with the "hope" that it works out. Lower wages and fewer benefits are simply easier to calculate. However, it may be worth the risk. When companies make their people feel like they matter, the people come together in a way that money simply cannot buy.

When Will Is Strong

His banker and his entrepreneur friends warned him not to do it. They told him that if the company went forward with the plan that the employees would hate it. "They will leave," his friends said. However, the CEO also spent time talking to various people within the company to get their input before making a decision. And they all agreed. The company should implement a salary freeze and stop matching 401(k)s.

During the 2008 recession, when people were tightening their belts in hard economic times, many chose to put off buying nonessential items, like storage and organization products for their homes and offices. And The Container Store, the only national retailer solely devoted to storage and organization products for our homes and offices, felt it. Their sales dropped 13 percent. This presented a problem for a company unaccustomed to dips in revenue. They had enjoyed a compounded annual growth rate of 20 percent from when they first opened their doors in 1978. Leadership talked to some of the employees and concluded they had to cut their expenses by at least the same amount as the drop in sales. To add to the stress, no one knew how long the recession would last or how long sales would continue to drop.

The Container Store has always prided itself on being an employee-first kind of company. So when the recession hit, they refused to take the expedient route and lay off employees. But they had to do something. As they presented the plan to freeze salaries and 401(k) matches for an undetermined time period, leadership wasn't sure what to expect in response. They hoped their people would be understanding and agree that it was better that they should all share the hardship than ask a few to suffer more.

What actually happened surprised and delighted them beyond their expectations. Something happened that they had neither requested nor demanded. Not only did the people accept the pay freezes, they also took it upon themselves to find more ways to help save money. Though not required to, people who traveled for business downgraded their hotels—opting for a Hampton Inn over a Hilton, for example. Some stayed with friends and family, foregoing hotels altogether. Others simply didn't submit expense reports, opting instead to pay for their own meals and taxis while away. Any and every place they could save money, they did. Employees also reached out to vendors to ask if they could find ways to save the company money too. Amazingly, the vendors were eager to help. That's practically unheard of! Clearly, they were under no obligation to trim their prices just to help a customer that was feeling the crunch of hard times. But because The Container Store had such strong relationships with their suppliers, they *wanted* to help.

"Top down couldn't have been even half as effective," says Kip Tindell, the company's cofounder and former CEO. And he's right. A company's leadership can demand that employees downgrade their hotels, pressure their people to insist that vendors find savings and announce that they will no longer reimburse business trip expenses. And if they did those things, they would indeed save money . . . and also risk inciting mass rebellion. Lesser demands have been known to stir up silent and seething anger toward companies and their leadership. At The Container Store, because the desire to contribute came from the people themselves, the outcome was quite different. There was an electricity in the air. Morale ran high. People were excited to find ways to help. Most important, everyone felt like they were in it together.

Very often, finite-minded leaders believe the source of will is externally motivated—pay packages, bonuses, perks or internal competition. If only that's all it took to inspire a human being. Money can buy a lot of things. Indeed, we can motivate people with money; we can pay them to work hard. But money can't buy true will. The difference between an organization where people are extrinsically rewarded to give their all and one where people are intrinsically motivated to do so is the difference between an organization filled with mercenaries versus one filled with zealots. Mercenaries work hard only so long as we keep paying top dollar for their effort. There is little loyalty to the company or the team. There is no real sense of belonging or feelings that anyone is contributing to something larger than themselves. Mercenaries are not likely to sacrifice out of love and devotion. In contrast, zealots *love* being a part of the organization. Though they may get rich doing what they are doing, they aren't doing it to get rich. They're doing it because they believe in the Just Cause.

At The Container Store, Tindell says, "Our employees put the cause before themselves." Though important, it was not the Just Cause alone, however, that inspired the will of the people. What Tindell saw during the recession was the payoff on a long-term investment. Tindell remembers what happened during the 2008 recession as a display of "spontaneous love and devotion." It may have felt spontaneous to him, but it wasn't. Strong will cannot be built overnight and it doesn't come from nothing. For years The Container Store had provided a great place to work, paid frontline employees better than most other retail jobs and trained leaders to put people's personal growth before the company's financial growth. And for years, their people had, in turn, taken care of their customers, the company and their vendors. And now, with the company in trouble, the people and the vendors wanted to do what was right by the company. How we treat people is how they treat us.

One reason companies that operate with a bias for will ultimately fare better in the Infinite Game has to do with what we can control. Though we have control over how we spend or manage our money, we have a lot less control over how we make it. Politics, economic cycles, market fluctuations, the actions of other players, customer preferences, technological advancements, the weather and all other forces majeures can wreak havoc with our ability to amass resources. Leaders can exert only limited control over any of these things. However, leaders have near total control over the source of will. Will is generated by the company culture.

Unlike resources, which are ultimately limited, we can generate an endless supply of will. For this reason, organizations that choose to operate with a bias for will are ultimately more resilient than those who prioritize resources. When hard times strike (and hard times always strike), in companies with a bias for will, the people are much more likely to rally together to protect each other, the company, the resources and their leaders. Not because they are told to, but because they choose to. This is what happens when the will of the people is strong. "We built a sense of family—of love and loyalty to each other, our customers, vendors and communities. Our intention was to build a business where everyone associated with it thrives," says Tindell.

Chapter 7

TRUSTING TEAMS

W hat is this for?" asked George. "This has nothing to do with the oil field." This was the general consensus from the rest of the people in the room too. They were to be the crew for the Shell URSA, the biggest offshore deepwater drilling platform the Shell Oil Company had ever built and they had no time for this "workshop."

The Shell URSA would stand forty-eight stories tall and would be capable of drilling deeper than any other platform in the world, more than three thousand feet below the surface of the ocean. At the time, 1997, it cost $1.45 billion to build (about $5.35 billion in today's dollars). Given how massive and expensive an operation it was, it presented all kinds of new challenges and dangers, so Shell wanted things done right. Which is why they handpicked Rick Fox as the man to lead the job.

Fox was a tough guy's tough guy. Hard and confident. He was intolerant of weakness. He felt he had every right to be. This was one of the most dangerous jobs in the world. One false step, a glance in the wrong direction and in an instant a man could be ripped in two and killed by one of the heavy moving parts. He knew so—he'd seen it happen. Safety was Fox's number one concern . . . that, and making sure that the URSA operated at peak capacity, pulling as many barrels of oil out of the ocean floor as it could handle.

Off in Northern California, far from Shell's New Orleans headquarters, lived a woman named Claire Nuer. A Holocaust survivor, Nuer operated a leadership consulting practice. She heard about the Shell URSA and, always looking for opportunities to share her philosophies, cold-called Rick Fox. When Nuer asked Fox about the challenges he faced, he spent most of the time telling her about the technical challenges. After letting him explain all the complexities of running a deep-sea rig, Nuer made a rather unusual proposal. If Fox *really* wanted his crew to be safe and succeed in the face of all the new challenges, his crew would need to learn to express their feelings.

Such an idea must have sounded ooey-gooey and New Agey. It must have sounded like it had no place in any serious, performance-driven organization. If it were any other time, Fox, a man who believed expressing feelings was the same as expressing weakness, might have hung up the phone. But Nuer got lucky. For some reason, perhaps because he was struggling with a strained relationship with his son, Fox listened to what she had to say. He even accepted an invitation to fly to California with his son to attend one of her workshops. There, father and his son were offered a safe space to open up about how they felt about each other. The workshop had such a profound and positive impact on their relationship that Fox wanted others to experience it too. He hired the Northern California, hippie type to fly across the country and test her theories with his roughneck, calloused, Louisiana crew. He knew they would be cynical and laugh at what he was asking them to do. But Fox cared about his crew, and he knew that any humiliation or mockery he would have to endure would be short lived compared to the benefit they would gain. And so the experiment began.

Day after day, for hours, members of the URSA crew would sit in circles and talk about their childhoods and their relationships. Their happy memories and their not-so-happy memories. On one occasion, a crew member broke down in tears as he told his teammates about his son's terminal illness. Crew members were not only asked to talk about themselves, there were also asked to listen. Another crew member recalled being prompted to ask the group, "If there was one thing you could change about me, what would it be?" "[You] don't listen," they told him, "you talk too much." To which he could only reply, "Tell me more."

The members of Fox's team got to know each other on a deeper level than ever before. Not just as coworkers but as humans. They opened up about who they were versus who they pretended to be. And as they did, it became clear that, for most of them, the tough-guy personas they projected were just that—personas. Under their hard exteriors, like all people, they had doubts, fears and insecurities. They had just been hiding them. Over the course of a year, Rick Fox, with Claire Nuer's guidance, built a team for the Shell URSA whose members felt psychologically safe with each other.

There is a difference between a group of people who work together and a group of people who trust each other. In a group of people who simply work together, relationships are mostly transactional, based on a mutual desire to get things done. This doesn't preclude us from liking the people we work with or even enjoying our jobs. But those things do not add up to a Trusting Team. Trust is a feeling. Just as it is impossible for a leader to demand that we be happy or inspired, a leader cannot order us to trust them or each other. For the feeling of trust to develop, we have to feel safe expressing ourselves first. We have to feel safe being vulnerable. That's right, vulnerable. Just reading the word makes some people squirm in their seats.

When we work on a Trusting Team we feel safe to express vulnerability. We feel safe to raise our hands and admit we made a mistake, be honest about shortfalls in performance, take responsibility for our behavior and ask for help. Asking for help is an example of an act that reveals vulnerability. However, when on a Trusting Team, we do so with the confidence that our boss or our colleagues will be there to support us. "Trust is the stacking and layering of small moments and reciprocal vulnerability over time," says Brené Brown, research professor at the University of Houston in her book *Dare to Lead*. "Trust and vulnerability grow together, and to betray one is to destroy both."

When we are not on a Trusting Team, when we do not feel like we can express any kind of vulnerability at work, we often feel forced to lie, hide and fake to compensate. We hide mistakes, we act as if we know what we're are doing (even when we don't) and we would never admit we need help for fear of humiliation, reprisal or finding ourselves on a short list at the next round of layoffs. Without Trusting Teams, all the cracks in an organization are hidden or ignored. Which, if that continues for any length of time, will compound and spread until things start to break. Trusting Teams, therefore, are essential to the smooth running of any organization. And on an oil rig, it actually saves lives.

"Part of safety," said Professor Robin Ely, coauthor of the *Harvard Business Review* article about the URSA, "is being able to admit mistakes and being open to learning—to say, 'I need help, I can't lift this thing by myself, I'm not sure how to read this meter.'" What the URSA crew discovered is that the more psychologically safe they felt around each other, the better information flowed. For the first time in many of their careers, Fox's crew felt safe to raise concerns. And the results were remarkable. The Shell URSA had one of the best safety records in the industry. And as Nuer's trust-building techniques spread across the company, it contributed to an 84 percent overall decline in accidents companywide.

When I suggest that teams must learn to be vulnerable with one another, that they must care about each other and show it, I often face pushback. The chief of a state police department, for example, told me: "I understand what you're saying, but I can't go back to my organization and tell the officers I 'care' about them. It's a machismo culture. I just can't do it. It won't work." But if a roughneck like Rick Fox can do it on an oil rig, then any leader in any industry can do the same. Our ability to trust is not based on our industry. This is human being stuff. Sometimes all we need to do is translate the concepts to fit the cultures in which we work. I asked the chief, "Can you go back to your officers and tell them, 'I give a shit about you guys. I want you to come to work and feel like I give a shit about you and I want to build a culture in which every officer feels like someone gives a shit about them'?" The chief smiled. He could do that.

In business, the resistance tends to come from a different place. Leaders of companies tell me that business is supposed to be professional, not personal. That their job is to drive performance, not to make their people feel good. But the fact is, there is no avoiding the existence of feelings. If you've ever felt frustrated, excited, angry, inspired, confused, a sense of camaraderie, envious, confident or insecure while at work, then congratulations, you're human. There is no way we can turn off our feelings simply because we are at work.

Feeling safe to express our feelings is not to be confused with a lack of emotional professionalism. Of course, we can't rage or disengage because we're feeling upset with someone on our team. We are still adults and we must still act with respect, courtesy and thoughtfulness. However, this does not mean we can or should even try to turn off our emotions. To deny the connection between feelings and performance is a finite-minded way of looking at leadership. In contrast, leaders like Rick Fox understand that feelings are at the heart of Trusting Teams . . . and Trusting Teams, it turns out, are the healthiest and highest-performing kind of teams.

On oil rigs, the historical average for industry uptime (the amount of time a platform is up running and operational) is 95 percent. The Shell URSA ran at 99 percent uptime. Their production was 43 percent better than industry benchmarks; they even outperformed their own production goals by 14 million barrels. And as if that weren't enough, the URSA was way ahead of their targets for environmental goals as well. In other words, to build high-performing teams, trust comes before the performance.

Performance vs. Trust

The Navy SEALs became well known to the public from movies like *Acts of Valor* and *Captain Phillips* and from the operation that resulted in the death of al-Qaeda leader Osama bin Laden. Indeed, the Naval Special Operations Forces are among the highest-performing organizations on the planet. However, it may surprise you to learn that the people on their teams are not necessarily the highest-performing individuals. To determine the kind of person who belongs in the SEALS, one of the things they do is evaluate candidates on two axes: performance versus trust.

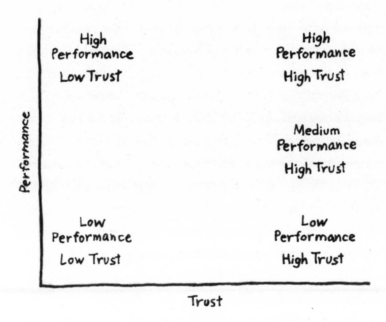

Performance is about technical competence. How good someone is at their job. Do they have grit? Can they remain cool under pressure? Trust is about character. Their humility and sense of personal accountability. How much they have the backs of their teammates when not in combat. And whether they are a positive influence on other team members. The way one SEAL team member put it, "I may trust you with my life but do I trust you with my money or my wife?" In other words, just because I trust your technical skills doesn't mean I think you are trustworthy as a person. You might be able to keep me safe in battle, but I don't trust you enough to be vulnerable with you personally. It's the difference between physical safety and psychological safety.

Looking at the Performance vs. Trust graph, it is clear that no one wants the person in the lower-left corner on their team, the low performer of low trust. Clearly, everyone wants the person in the top-right corner on their team, the high performer of high trust. What the SEALs discovered is that the person in the top left of the graph—the high performer of low trust—is a toxic team member. These team members exhibit traits of narcissism, are quick to blame others, put themselves first, "talk shit about others" and can have a negative influence on their teammates, especially new or junior members of the team. The SEALs would rather have a medium performer of high trust, sometimes even a low performer of high trust (it's a relative scale), on their team than the high performer of low trust. If the SEALs, who are some of the highest-performing teams in the world, prioritize trust before performance, then why do we still think performance matters first in business?

In a culture dominated by intense pressure to meet quarterly or annual targets, too many of our leaders value high performers with little consideration of whether others on the team can trust them. And those values are reflected in whom they hire, promote and fire. Jack Welch, CEO of GE during much of the high-flying 1980s and '90s, offers an extreme example of what this looks like. Welch was so concerned with winning and being number one (he even titled one of his books *Winning*) that he focused almost exclusively on performance to the detriment of trust. Like the SEALs, Welch also ranked his executives on two axes. Unlike the SEALs, however, his axes were performance and potential; basically, performance and future performance. Based on these metrics, those who "won" biggest in a given year were earmarked for promotion. The underperformers were fired. In his drive to produce a high-performing culture, Welch focused on someone's output above all else. (Though Welch did have metrics on culture, if you ask anyone who worked at GE at the time, it was largely ignored.)

Environments like the one Welch cultivated tend to benefit and celebrate the high performers, including the ones of low trust. The problem is, the toxic team members are often more interested in their own performance and career trajectories than they are with helping the whole team rise. And though they may crush it in the near term, the manner in which they achieve their results will often contribute to a toxic environment in which others will struggle to thrive. Indeed, in performance-obsessed cultures, these tendencies are often exacerbated by leaders who encourage internal competition as a way to further drive performance.

Pitting their people against each other might seem like a good idea to finite-minded leaders like Welch. But it's only good for now. Eventually, it can lead to behaviors that actually undermine trust, things like hoarding information instead of sharing it, stealing credit instead of giving it, manipulating younger team members and throwing others under the bus to avoid personal accountability. In some cases, people will go so far as to intentionally sabotage their colleagues to advance themselves. As expected, in time, the organization as a whole will suffer . . . maybe to the point that it is forced out of the game altogether. The GE that Jack built was almost destined to fail before too long. Indeed, if it weren't for a $300 billion government bailout they received after the 2008 stock market crash, GE probably wouldn't exist anymore. Time is always the great revealer of truth.

It's not surprising that even well-intentioned leaders who value trust often fall into the trap of hiring and promoting high performers without regard to whether they can be trusted and trusting. Performance can easily be quantified in terms of output. Indeed, in business, we have all sorts of metrics to measure someone's performance, but we have few if any effective metrics to measure someone's trustworthiness. The funny thing is, it is actually incredibly easy to identify the high performers of low trust on any team. Simply go to the people on the team and ask them who the asshole is. They will likely all point to the same person.

Conversely, if we ask team members whom they trust more than anyone else on the team, who is always there for them when the chips are down, they will likely also all point to the same person. That person may or may not be the highest individual performer, but they are a great teammate and may be a good natural leader, able to help raise the group's performance. These team members tend to have a high EQ and take personal accountability for how their actions affect the team's dynamics. They want to grow and help those around them grow too. Because we tend to measure only someone's performance and not trust, we are more likely to miss the value of a trusted team member when deciding whom to promote.

When confronted with the information about how others feel about them, high performers of low trust rarely agree or even want to listen. They think of themselves as trustworthy, it's everyone else who can't be trusted. They offer excuses instead of taking responsibility for how they show up. And though they can feel that the rest of the team may not include them in things (probably convincing themselves that everyone is jealous of them), they fail to recognize that the only common factor in all these tense relationships is them. Even when told how the rest of the team feels about them, many higher performers of low trust will double down on performance instead of trying to repair lost trust. After all, thanks to lopsided corporate metrics, it is their performance that helped them advance their careers and provide job security in the past. Why change strategy now?

Good leaders don't automatically favor low performers of high trust nor do they immediately dump high performers of low trust. If someone's performance is struggling or if they are acting in a way that is negatively impacting team dynamics, the primary question a leader needs to ask is, "Are they coachable?" Our goal, as leaders, is to ensure that our people have the skills—technical skills, human skills or leadership skills—so that they are equipped to work to their natural best and be a valuable asset to the team. This means we have to work with the low-trust players to help them learn the human skills to become more trusted and trusting, and work with the low performers to help them learn the technical skills to improve their performance. Only when a team member proves uncoachable—is resistant to feedback and takes no responsibility for how they show up at work—should we seriously consider removing them from the team. And at that point, should a leader still decide to keep them, the leader is now responsible for the consequences.

Teams naturally ostracize or keep at arm's length the member they don't trust. The one who "is not one of us." This should make it easier for a leader to know whom to coach or remove so that the whole team's performance can rise. Or does it? Is it the team member who is low trust or is it the rest of the team?

If You Build It, They Will Come

There had been several allegations made against him. Investigators were looking into some of them, including whether he was sleeping in the gym instead of being out on patrol, whether he had illegally tinted windows on his personal vehicle and whether he tried to use his badge to get out of a ticket in another jurisdiction. He was even being investigated for having sex with his ex-wife in a patrol car while on duty. Officer Jake Coyle felt like they were constantly going after him for something. Like the microscope was always on him. He didn't trust his leaders, he didn't trust his colleagues and they didn't trust him.

Other police officers regularly picked on Officer Coyle. He wasn't a member of their clique and they made sure he knew it. They made fun of him and played pranks on him. They would put garbage in his car, for example, or block his car in with a snowplow. To the other officers, it was just playful hijinks, frat-boy humor. But to Officer Coyle it was much more serious. Their behavior toward him left him feeling no sense of trust or psychological safety within the department. It got to the point that he hated coming to work. He just wanted to get through his shift and go home. More and more, he was thinking about picking up and starting over somewhere else; he was already looking into a transfer to a different police department. And then something happened.

When Jack Cauley arrived at the Castle Rock Police Department to be the new chief, what he found was a police force that resembled the one he had just left and countless others around the country (as well as too many corporate cultures today). A place where many people felt undervalued and ignored. Where they felt pressured to make the numbers above all else. "We were basically told that we were replaceable and that there [were] hundreds of people waiting to have our jobs," said one officer, describing what it was like at CRPD before Chief Cauley. "Rookies [did] not feel comfortable advancing ideas they may have [had]," another said. It was a place where officers would be punished for not writing enough tickets.

Chief Cauley knew all about police departments using tickets and arrests as the only metrics of performance. As an ambitious young officer starting his career in Overland Park, Kansas, in 1986, he himself had climbed the ranks by beating the metrics his superiors set for him. If they wanted him to write X many tickets, he would write double. Over the years, he came to realize that such a focus on performance came at a cost to the officers and the culture of policing. So, when he was offered the job to be the chief at Castle Rock PD, he leapt at it. This was his chance to prove what can happen to a police force with a culture built on trust, not tickets written, blind obedience or job insecurity.

One of Cauley's first acts as police chief was to hold listening sessions with every single member of the organization—every sworn officer and every staff member. During the sessions, multiple people told him that they had been asking for years for a fence to be built around the parking lot. The parking lot was an open and exposed area of asphalt that wrapped around the CRPD headquarters. Officers and staff complained that when they left work at night, when it was quiet and dark outside, they felt afraid walking to their cars. They had no idea if someone was hiding, waiting to pounce on them. For years, management told them to deal with it. They were told that there were more pressing things to spend money on than a fence, things more related to the job of policing—like new firearms or new cars.

It became clear to Cauley that the people who worked at the department did not feel like their leaders had their back. The new chief had to build a "Circle of Safety" first. Without it, nothing else he needed to do would work. So, in short order, Cauley had a fence erected around the parking lot. This simple act put everyone on notice: things were going to change. It was one of a series of seemingly small things that sent a profound message to his people—I hear you and I care about you. A Circle of Safety is a necessary condition for trust to exist. It describes an environment in which people feel psychologically safe to be vulnerable around their colleagues. Safe to admit mistakes, point out gaps in their training, share their fears and anxieties and, of course, ask for help with the confidence that others will support them instead of using that information against them.

It was during one of his early listening sessions that Cauley sat down with "Problem Officer" Jake Coyle. The chief knew that internal investigations had exonerated Coyle from the more significant allegations against him. A few infractions, however, proved true, like having illegally tinted windows on his personal vehicle. None of the violations were major, but together they were enough to fire the young officer. Chief Cauley could have looked at Officer Coyle, said, "Low performer, low trust," and shown him the door. But Chief Cauley suspected that it was the culture that was toxic, not the officer. And if he was working to change the culture, then it only seemed fitting that he give the officer a second chance. To many a finite-minded leader, the chief's decision would be considered too risky; why keep an employee who has proved themselves to be a lower performer and untrustworthy? Instead of terminating Jake Coyle, however, Chief Cauley gave him a three-day unpaid suspension and, as Coyle remembers the chief telling him, "the opportunity to turn this around." Officer Coyle smiles as he tells the rest of the story. "He basically said 'I believe in you. . . .' [My job] was basically the one thing I had left. I already lost everything else . . . and so I was like, 'Okay. Let's do this!'"

With those words Officer Coyle showed that he knew he had work to do. If his chief wanted to build a culture of trust, then he had to act in a way that would be worthy of that trust. True trusting relationships require both parties to take a risk. Like dating or making friends, though one person has to take a first risk to trust, the other person has to reciprocate at some point if the relationship has any chance of succeeding. In an organization, it is the leader's responsibility to take the first risk, to build a Circle of Safety. But then it is up to the employee to take a chance and step into the Circle of Safety. A leader cannot force anyone into the circle. Even on strong, Trusting Teams there are still some who refuse to step in, especially on teams with an entrenched history of prioritizing performance before trust. This does not mean they are toxic, it just means they need more time. True trust takes time to develop and it can take some people longer than others.

The process of building trust takes risk. We start by taking small risks, and if we feel safe, we take bigger risks. Sometimes there are missteps. Then we try again. Until, eventually, we feel we can be completely ourselves. Trust must be continuously and actively cultivated. For Chief Cauley, giving Officer Coyle a second chance to make something of himself in a healthier culture was just the start. He stayed personally involved in Coyle's growth. He coached him now and then, checked in on him every so often and kept tabs on how he felt about his job, and made sure that Officer Coyle's direct supervisors were doing the same thing. Chief Cauley also held Coyle accountable for his own actions and offered him a safe space to express how he felt without any fear of humiliation, taunting or retribution. Coyle, in turn, had to take advantage of the safe space Cauley was building to share his feelings and ask for help when he needed it. He was also expected to behave in a way that was consistent with the values of the organization. And it worked. Today, the culture of the Castle Rock Police Department has been completely transformed. It is a place in which trust flows freely. Jake Coyle is now one of the most respected and most trusted officers at CRPD and is responsible for training new recruits who join their ranks. And Chief Cauley, always in search of the truth, still does his listening sessions.

The Truth Shouldn't Hurt

Human beings are hardwired to protect ourselves. We avoid danger and seek out places in which we feel safe. The best place to be is among others around whom we feel safe and who we know will help protect us. The most anxiety-inducing place to be is alone—where we feel we have to protect ourselves from the people on our own team. Real or perceived, when there is danger, we act from a place of fear rather than confidence. So just imagine how people act when they work in constant fear of missing out on a promotion, fear of getting in trouble, fear of being mocked, fear of not fitting in, fear of their boss thinking they're an idiot, fear of finding themselves on a short list for the next round of layoffs.

Fear is such a powerful motivator that it can force us to act in ways that are completely counter to our own or our organization's best interests. Fear can push us to choose the best finite option at the risk of doing infinite damage. And in the face of fear, we hide the truth. Which is pretty bad in any circumstance, but when an organization is doing badly, it's even worse. This is exactly what Alan Mulally walked into when he took over as the new CEO at Ford in 2006.

Ford was in serious trouble, and Mulally was brought in with the hope that he could save the company. Much as Chief Cauley had done at the CRPD, Mulally made it his first order of business at Ford to find out as much as he could about the current state of things from the people who worked there. The task, however, proved more difficult than he expected.

To keep a pulse on the health of the organization, Mulally introduced weekly business plan reviews (BPRs). All his senior executives were to attend these meetings and present the status of their work against the company's strategic plan, using simple color coding—green, yellow and red. Mulally knew that the company was having serious problems, so he was surprised to see that week after week every executive presented their projects as all green. Finally, he threw up his hands in frustration. "We are going to lose billions of dollars this year," he said. "Is there anything that's *not* going well here?" Nobody answered.

There was a good reason for the silence. The executives were scared. Prior to Mulally, the former CEO would regularly berate, humiliate or fire people who told him things he didn't want to hear. And, because we get the behavior we reward, executives were now conditioned to hide problem areas or missed financial targets to protect themselves from the CEO. It didn't matter that Mulally said he wanted honesty and accountability; until the executives felt safe, he wasn't going to get it. (For all the cynics who say there is no place for feelings at work, here was a roomful of the most senior people of a major corporation who didn't want to tell the truth to the CEO because of how they felt.) But Mulally persisted.

In every subsequent meeting he repeated the same question until, eventually, one person, Mark Fields, head of operations in the Americas, changed one slide in his presentation to red. A decision he believed would cost him his job. But he didn't lose his job. Nor was he publicly shamed. Instead, Mulally clapped at the sight and said, "Mark, that is great visibility! Who can help Mark with this?" At the next meeting, Mark was still the only executive with a red slide in his presentation. In fact, the other executives were surprised to see that Fields still had his job. Week after week, Mulally would repeat his question, We are still losing tons of money, is anything *not* going well? Slowly executives started to show yellow and red in their presentations too. Eventually, it got to the point where they would openly discuss all the issues they were facing. In the process, Mulally had learned some tricks to help build trust on the team. To help them feel safe from humiliation, for example, he depersonalized the problems his executives faced. "You *have* a problem," he would tell them. "*You* are not the problem."

As the slides presented at the BPR meetings became more colorful, Mulally could finally see what was actually going on inside the company, which meant he could actively work to give his people the support they needed. Once the Circle of Safety had been established, a Trusting Team formed and the executives could now, in Mulally's words, "work together as a team to turn the reds to yellow and the yellows to green." And if they could do that, he knew they could save the company.

Nothing and no one can perform at 100 percent forever. If we cannot be honest with one another and rely on one another for help during the challenging parts of the journey, we won't get very far. But it's not enough for leaders to simply create an environment that is safe for telling the truth. We must model the behavior we want to see, actively incentivize the kinds of behaviors that build trust and give people responsible freedom and the support they need to flourish in their jobs. It is the combination of what we value and how we act that sets the culture of the company.

Culture = Values + Behavior

To build a culture based on trust takes a lot of work. It starts by creating a space in which people feel safe and comfortable to be themselves. We have to change our mindset to recognize that we need metrics for trust and performance before we can assess someone's value on a team. This is perhaps one of the most powerful components of Chief Cauley's transformation of the Castle Rock Police Department. A culture in which pressure to meet numbers was replaced with a drive to take care of one another and serve the community better. To do this, however, he knew that he would need to change the way that he recognized and rewarded his people.

These days, CRPD officers' evaluations focus on the problems they are solving and the impact they are making in the lives of people at the department and in the community. The traditional metrics are included, but they aren't the focus any more. In addition to written evaluations, Cauley also occasionally presents certificates of recognition during roll call. These go to the officer or officers whose work best embodies the values of the department.

Unsurprisingly, because Chief Cauley promotes and recognizes care for team members and community, initiative and problem solving over traditional metrics, what he gets is more care, more initiative and more problem solving. Again, we get the behavior we reward. And the more problems the people of the Castle Rock Police Department solve, the more initiative they show, the more trust has flourished in the force and with the community. Chief Cauley calls it "one-by-one policing," because the benefits build up one step, one problem solved at a time. It's a system that promotes consistency over intensity.

People will trust their leaders when their leaders do the things that make them feel psychologically safe. This means giving them discretion in how they do the jobs they've been trained to do. To allow people to exercise responsible freedom. Whereas in the old system they were told, "Go do A, B, C, D and repeat," explains Chief Cauley, in the new system, when officers saw a problem or opportunity and said, "Wouldn't it be cool if . . . ," Chief Cauley let them run with it.

This is the core of one-by-one policing. Good leadership and Trusting Teams allow the people on those teams to do the best job they can do. The result is a culture of solving problems rather than putting Band-Aids on them. It's the difference between issuing lots of tickets at an intersection that has a lot of accidents and figuring out how to reduce the number of accidents in the first place. It also deters overzealous policing that can come as a result of a lopsided, metrics-heavy system of evaluation and recognition.

The bicycle unit, for example, knew about an unused bike track in town and saw an opportunity. They took the initiative to put the word out that any kids with bicycles were invited to come learn to jump their bikes, ride on the track and have free doughnuts with the officers—Dirt, Jumps and Doughnuts, they called it. The officers showed up with doughnuts donated by a local shop, a table, their bicycles and waited. The first time they did it, they expected few kids to show up. In fact, over forty kids showed up, a number that has remained consistent every single month. Dirt, Jumps and Doughnuts became a huge opportunity for community engagement. For most people, the only time we talk to a police officer is if something has gone wrong or if we are trying to get ourselves out of trouble. These officers wanted to get to know the kids and they wanted the kids to get to know them beyond a one-time show-and-tell at the local school, for instance. At Dirt, Jumps and Doughnuts, there are no presentations or formal requests made by the police, they just ride their bikes with the kids.

On one occasion, the department received a call that a resident believed the house next to theirs was being used to sell drugs. Traditionally in such cases, the police would initiate an investigation. This would often be done covertly and include undercover officers both surveilling the house or making a buy. All the while, the neighbor who made the call wouldn't see a police response and would feel ignored. After weeks or months of building a case, the police would obtain a warrant, gather a larger group of heavily armed officers and forcibly break down the door to raid the house. The practice is dangerous for everyone involved, and though some arrests may be made, as officers explained to me, before long "[the dealers] would often be back on the streets and maybe back in that same house back at it." And even if the officers are successful in shutting down the house, the crime scene is often left wrapped in police tape with the doors broken in— not exactly something other neighbors want to be left with.

The new culture at CRPD opened up the opportunity to try something different. Instead of a stakeout, one of the officers walked up to the alleged drug house and knocked on the front door. When a person answered, the officer didn't ask to enter; instead, they shared that there had been reports about possible drug deals at the house and informed the person inside that the police would be watching. Over the next few weeks, the police presence in the area was stepped up. Officers on their rounds would make a point to drive by the house, maybe park across the street to eat their lunch. As it turns out, it's very hard to sell drugs from a house in which there is a regular police presence outside. And so the tenants simply left. No doors bashed down. No lives put at risk.

Now I fully appreciate the cynical view of this. That the police didn't solve the problem, they simply moved it to another location. And now another jurisdiction would have to deal with the problem and risk their lives. I grant you that this is indeed the case. But this is an infinite game. Using this one-by-one system of policing, the aim would be for other departments to adopt similar tactics and further develop their own. In time, a crime like selling drugs out of neighborhood

homes becomes a more difficult business proposition altogether, city by city, state by state, one by one. Notice that I
said "more difficult" and not impossible. Despite what we've
been led to believe by those who talk about the "war on
drugs," this is not a game that can be won. Drug dealers
aren't trying to beat the police and win; they are just trying to
do more drug deals. The police need to play with the right
mindset for the game they are in.

Infinite games, remember, require infinite strategies. Because crime is an infinite game, the approach Chief Cauley's
officers are taking is much better suited to that game than an
attack-and-conquer mindset. The goal is not to win in the
overall scheme of things; the objective is to keep your will
and resources strong while working to frustrate the will and
exhaust the resources of the other players. Police can never
"beat" crime. Instead, the police can make it more difficult
for the criminals to be criminals. At CRPD Chief Cauley's
officers are developing strategies that can be easily, cheaply
and safely repeated over and over . . . forever if necessary.

"Most of what cops do is address quality of life issues, not
fighting crime," explains Chief Cauley, "and what about the
quality of life for the officers?" If someone has to muster the
energy to go to a job they hate every day, it will take a toll on
their confidence and negatively affect their judgment. "If a
cop's grumpy, you're probably screwed," one officer explained.
"If he's having a bad day and you're making it even worse for
him or making more work for him, you're probably going to
get the worst of it." Just like the Shell URSA, when a job can
be deadly, creating a space in which employees can feel safe to
open up is more than a nice-to-have, it's essential.

If an officer feels inspired to go to work every day, feels
trusted and trusting when they are there and has a safe and

healthy place to express their feelings, the odds are pretty high that members of the public who interact with that officer will benefit too. Just as customers will never love a company until the employees love the company first, the community will never trust the police until the police trust each other and their leaders first.

Adding new focus on the culture inside the organization as a way to address outside challenges, the Castle Rock Police Department has seen a remarkable shift among its 75 sworn officers. Considering that over 95 percent of the nearly 12,500 police departments in the United States have fewer than 100 officers, one-by-one policing could serve as a model for other police departments that may be struggling with trust issues inside the department or with the community.

Indeed, Chief Cauley recognizes that there is still a lot to do in his own department and that the old way of thinking hasn't completely gone away. But CRPD is on a journey and their culture today is significantly healthier than it used to be. Anecdotally, the officers report a significant increase in the number of people in the community who will wave them down just to say thank you. They report significantly more people buying them cups of coffee at coffee shops. Crime is under control and the community is more willing to help out too. "The community sees us as problem solvers," says Chief Cauley, "not the enforcers."

If leaders, in any profession, place an excess of stress on people to make the numbers, and offer lopsided incentive structures, we risk creating an environment in which near-term performance and resources are prioritized while long-term performance, trust, psychological safety and the will of the people decline. It's true in policing and it's true in business. If someone who works in customer service is highly stressed at work, it increases the likelihood that they will provide a poor customer service experience. How they feel affects how they do their job. No news there. Any work environment in which people feel like they need to lie, hide and fake about their anxieties, mistakes or gaps in training for fear of getting in trouble, humiliated or losing their job undermines the very things that allow people to build trust. In the policing profession the impact can be much more serious than poor customer service.

In weak cultures, people find safety in the rules. This is why we get bureaucrats. They believe a strict adherence to the rules provides them with job security. And in the process, they do damage to the trust inside and outside the organization. In strong cultures, people find safety in relationships. Strong relationships are the foundation of high-performing teams. And all high-performing teams start with trust.

In the Infinite Game, however, we need more than strong, trusting, high-performing teams today. We need a system that will ensure that that trust and that performance can endure over time. If leaders are responsible for creating the environment that fosters trust, then are we building a bench of leaders who know how do to that?

How to Train a Leader

Would-be leaders in the U.S. Marine Corps attend a ten-week training and selection process at Officer Candidate School in Quantico, Virginia. Among the many tests administered at OCS is the Leadership Reaction Course. The LRC is a series of twenty mini obstacle courses—problem-solving courses, to be more accurate. Working in groups of four, the Marines are given challenges such as figuring out how to get all their people and matériel across a water hazard (military-speak for a pond) within a set time period using just three planks of different sizes. The Marine Corps uses the LRC to evaluate the leadership qualities of their future officers. They look at things like how well the candidates follow a leader or deal with adversity and how quickly they can understand a situation and prioritize and delegate tasks. The amazing thing is, of all the qualities those future leaders are assessed on, the ability to successfully complete the obstacle is not one of them. There isn't even a box to check at the bottom of the evaluation form. In other words, the Marine Corps focuses on assessing the inputs, the behaviors, rather than the outcomes. And for good reason. They know that good leaders sometimes suffer mission failure and bad leaders sometimes enjoy mission success. The ability to succeed is not what makes someone a leader. Exhibiting the qualities of leadership is what makes someone an effective leader. Qualities like honesty, integrity, courage, resiliency, perseverance, judgment and decisiveness, as the Marines have learned after years of trial and error, are more likely to engender the kind of trust and cooperation that, over the course of time, increase the likelihood that a team will succeed more often than it fails. A bias for will before resources, trust before performance, increases the probability a team will perform at higher levels over time.

The ability for any organization to build new leaders is very important. Think of an organization like a plant. No matter how strong it is, no matter how tall it grows, if it cannot make new seeds, if it is unable to produce new leaders, then its ability to thrive for generations beyond is nil. One of the primary jobs of any leader is to make new leaders. To help grow the kind of leaders who know how to build organizations equipped for the Infinite Game. However, if the current leaders are more focused on making their plant as big as possible, then, like a weed, it will do whatever it needs to do to grow. Regardless of the impact it has on the garden (or even the long-term prospects of the plant itself).

I know many people who sit at the highest levels of organizations who are not leaders. They may hold rank, and we may do as they tell us because they have authority over us, but that does not mean we trust them or that we would follow them. There are others who may hold no formal rank or authority, but they have taken the risk to care for their people. They are able to create a space in which we can be ourselves and feel safe sharing what's on our mind. We trust those people, we would follow them anywhere and we willingly go the extra mile for them, not because we have to, but because we want to.

The Marine Corps isn't interested in whether or not leaders can cross a water hazard or any other arbitrary obstacle. They are interested in training leaders who can create an environment in which everyone feels trusted and trusting so that they can work together to overcome any obstacle. Marines know that a leadership climate based on trust is what helps ensure they will enjoy success more often than not.

It's a phrase I will repeat again in this book: leaders are not responsible for the results, leaders are responsible for the people who are responsible for the results. And the best way to drive performance in an organization is to create an environment in which information can flow freely, mistakes can be highlighted and help can be offered and received. In short, an environment in which people feel safe among their own. This is the responsibility of a leader.

This is what Rick Fox did. He built a high-performing team by creating an environment in which his crew felt safe to be vulnerable around each other. The SEALs do this. They build high-performing teams by prioritizing an individual's trustworthiness over their ability to perform. Alan Mulally did this. He helped Ford become a high-performing company again only after he created a safe space for his people to tell the truth about what was going on. And this is what Jack Cauley is doing . . . and the results have been transformative. When leaders are willing to prioritize trust over performance, performance almost always follows. However, when leaders have laser-focus on performance above all else, the culture inevitably suffers.

Chapter 8

ETHICAL FADING

I t's hard to imagine that this actually happened. It is so far from ethical in any way. It's hard to imagine that a group of people, who I'm sure consider themselves good and honest, were able to behave in ways that, by any standard, are just plain wrong.

From mid-2011 to about mid-2016, employees at Wells Fargo Bank opened over three and a half million fake bank accounts. As *The New York Times* reported in 2016, "Some customers noticed the deception when they were charged unexpected fees, received credit or debit cards in the mail that they did not request, or started hearing from debt collectors about accounts they did not recognize. But most of the sham accounts went unnoticed, as employees would routinely close them shortly after opening them."

Ultimately, 5,300 Wells Fargo employees were fired as a result of their involvement in these deceptive practices. Practices that then CEO John Stumpf told Congress "go against everything regarding our core principles, our ethics and our culture." In a statement made to the press, the company echoed Stumpf, saying that "the vast majority of our team members do the right thing, every day, on behalf of our customers. . . . And if any of these things transpired, it's distressing and it's not who Wells Fargo is." In other words, Wells Fargo executives would like us to believe that the offenders were just a few bad apples. However, this was not an isolated act of a small group of people; this was the result of thousands of people acting over the course of years! The more likely scenario was that Wells Fargo's culture suffered from a severe case of ethical fading.

Ethical fading is a condition in a culture that allows people to act in unethical ways in order to advance their own interests, often at the expense of others, while falsely believing that they have not compromised their own moral principles. Ethical fading often starts with small, seemingly innocuous transgressions that, when left unchecked, continue to grow and compound.

While ethical lapses can happen anywhere, organizations run with a finite mindset are especially susceptible to ethical fading. As discussed in the previous chapters, cultures that place excessive focus on quarterly or annual financial performance can put intense pressure on people to cut corners, bend rules and make other questionable decisions in order to hit the targets set for them. Unfortunately, those who behaved dubiously but hit their targets are rewarded, which sends a clear message about the organization's priorities. Indeed, the reward systems in these organizations work to incentivize such behaviors. Those who meet their goals are given bonuses or promoted often without consideration of the manner in which they met their goals, while those who acted with integrity but missed their targets are penalized by being overlooked for recognition or advancement. This sends a message to everyone else in the organization that making the numbers is more important than acting ethically. Those who may have been loath to follow the unethical examples set by their colleagues succumb to the pressure as they start to feel it is the only way for them to get a bonus, get ahead or even protect their job. They will lose perspective and rationalize their ethical transgressions. "I gotta put food on the table," "It's what management wants," "I have no choice," and my personal favorite, "It's the industry standard," are all rationalizations we tell ourselves or tell others to help us mitigate any sense of guilt or responsibility we may feel.

As human beings we are blessed and cursed with our ability for rational thought. We try to make sense of the world around us. We can understand complex equations and we have the ability to be introspective. It is with our capacity for rational and analytical thought that we can think through hard problems and advance technology. We can also use this capacity for analytical thinking to explain or justify our behavior when we know it violates some deep-seated code of ethics or helps us avoid some sense of guilt we may harbor for a decision or action we took. It's like stealing something from a rich friend and saying to yourself, "They won't even notice. Besides, they can afford another one." We can rationalize it any way we want; we still stole something from our friend. When such rationalizations become commonplace inside an organization, the snowball grows and grows until unethical behavior pervades the entire organization and, in extreme cases, leads to the kind of corruption that happened at Wells Fargo.

A Culture of Pressure, Demands and Incentives

In 1973, two Princeton University psychology professors, John M. Darley and C. Daniel Batson, conducted an experiment to better understand how situational variables can affect our ethics, specifically, how pressure impacts our will to help someone in distress. They asked a group of seminary students to travel across campus to give a talk about the story of the Good Samaritan. The Good Samaritan is a parable from the New Testament in which a Samaritan, traveling from Jericho to Jerusalem, is the only person to stop to help a man who had been beaten, robbed and left on the side of the road.

To recreate the scene, the professors hired an actor to lie in an alley, slumped over like he had been mugged or hurt in some way. The students would have to pass him as they made their way across campus. Each time the experiment was conducted with a different group of students, the professors added a little bit of pressure to see how it would affect the students' behavior. One group had a lot of pressure to hurry across campus. "You're late," the experimenters told them. "They were expecting you a few minutes ago. We'd better get moving. The assistant should be waiting for you so you'd better hurry. It shouldn't take but just a minute." A second group had intermediate pressure put on them. "The assistant is ready for you, so please go right over." And the final group had only slight pressure added to them. "It'll be a few minutes before they're ready for you, but you might as well head on over. If you have to wait over there, it shouldn't be long."

When there was low pressure, 63 percent of the students stopped to help the injured man. With medium pressure, 45 percent stopped to lend assistance. And under high pressure, only 10 percent of the students stopped to help someone in apparent distress. Some even stepped right over him. The conclusion was stark. The students were good people who cared about service. They were all studying to be priests, for heaven's sake. However, when pressure was placed upon them, in this case time pressure, their will to do the right thing gave way to demands placed upon them. And it was under extremely high-pressure conditions that the people in the sales department at Wells Fargo were forced to operate.

Though there were plenty of positive reinforcements offered to those who were able to make their numbers, regardless of how they made them, there was also a sense of fear instilled in those who didn't. Some employees recall being pushed to sell anywhere from eight to twenty different products a day, and when they fell short of their goals their managers berated them. One employee remembered her manager telling her, "If you don't meet your solutions you're not a team player. If you're bringing down the team then you will be fired and it will be on your permanent record." The employee told her supervisors that she felt there was no ethical way she could meet their expectations and called the bank's ethics hotline multiple times to say as much. This is the kind of response we would hope for or expect from an employee when there is evidence of ethical lapses inside a large organization. But in the end, Wells Fargo decided to fire her rather than respond to her concerns. The expectation at Wells Fargo was that employees would never say when the quotas were impossible to meet; they were simply expected to find a way to meet them, whatever it took. As another Wells Fargo employee confessed, "It was the norm to just open sales unethically. It was what we were taught and we just did it."

Ethical fading is not an event. It doesn't just suddenly arrive like a switch was flipped. It's more like an infection that festers over time. Investigations into the scandal at Wells Fargo discovered an internal review from a decade before the scandal broke that revealed that the organization's toxic conditions and unethical behavior had already been identified. That original review concluded that there was "an incentive to cheat" based on fear of job loss. Though the results of the review were sent to the company's chief auditor, HR representatives and others, the leadership did nothing to correct it. In addition, by 2010, a year before the fake account practices began, there were a reported seven hundred whistleblower complaints about the questionable sales tactics at the company (the board of directors reported they knew nothing about them). John Stumpf became aware that his company had systemic problems as early as 2013. A 2017 board report revealed, however, that he knew of individual issues as far back as 2002, nearly fifteen years before the scandal! The same 2017 report charged that Carrie Tolstedt, former head of Wells Fargo Community Banks, not only knew about the wrongful sales practices, but actually "reinforced the high-pressure sales culture." She was also, according to the report, "notoriously resistant to outside intervention and oversight" and, along with others in leadership, "challenged and resisted scrutiny." One can only surmise that either she was subject to similar pressures and was afraid to speak out or she was rewarded handsomely for the results her department achieved.

Despite Wells Fargo's public statements that the scandal was confined to the retail sales group and that the majority of the company "does the right thing," there was plenty of evidence that the ethical fading ran wide and deep throughout the company. Overlapping with the timing of the fake account scandal, for example, the bank was also misrepresenting the quality of loans they sold. In 2018, the bank was fined $2.09 billion to settle that issue. The auto division of the bank also agreed to repay $80 million back to customers for selling them auto insurance they didn't sign up for. And the wholesale division, the group that Tim Sloan ran before he replaced John Stumpf to become CEO, fell under scrutiny for other ethical lapses that may have included money laundering.

Wells Fargo did eventually accept accountability for opening up those millions of fake accounts and was fined a total of $185 million for doing so. The punishment they received, putting aside the temporary embarrassment and short-term impact on their stock price, however, was barely a slap on the wrist. To put things in perspective, $185 million represents less than 1 percent of Wells Fargo's total profit of $22 billion the year they were fined and only 0.2 percent of their total revenues of nearly $95 billion. It's the equivalent of someone who makes $75,000 in annual salary being fined $150. Not much of a punishment.

None of the company's leaders was held criminally liable for allowing a culture in which their own people committed fraud (which is a crime) to exist. No one went to jail. There wasn't even a single indictment. Indeed, John Stumpf did lose his job and $41 million of unvested equity, but he was only fired as a response to massive public pressure. What's more, he walked away with over $134 million in pension accounts and stock. So not only can leaders who oversee cultures in which ethical fading happens go unpunished, they can actually profit from it . . . which incentivizes leaders to maintain the status quo. I personally find it quite troubling when executives take credit for their "culture of performance," yet take no responsibility for a culture consumed by ethical fading.

When Good People Do Bad Things

As anyone who suffers from a life-threatening allergy to peanuts, bees or shellfish well knows, a shot of epinephrine can save your life. And given its 90 percent market share, the odds are high that you'll get that shot from an EpiPen. The EpiPen is a brand of epinephrine autoinjector that stops anaphylactic shock. The product is essential for anyone with an extreme allergy, and because it has a twelve-month life span, it has to be replaced on an annual basis. And at a cost of one hundred dollars for a two-pack, that makes for a good business.

In 2007, a company called Mylan bought the rights to the EpiPen brand. Given the dominance the brand had on the market, combined with the fact that there was no generic option at the time, there was nothing to stop Mylan from raising the price of the product by an average of 22 percent per year. Seeing the impact these price increases had on their stock value, in 2014 the board decided to up the ante. They offered select employees a one-time opportunity. If they doubled the company's earnings per share over the next five years, they would share in what could be hundreds of millions of dollars in bonuses. The top five executives alone would stand to make nearly $100 million. No doubt responding to this incentive, in the following year the company sped up the rate of EpiPen price increases from 22 percent to 32 percent. After the fifteenth price hike since 2009, in 2016 Mylan announced that a pair of EpiPens would now cost an all-time high of six hundred dollars, representing a 500 percent increase over just six years. The company probably would have continued to raise the price had it not been for a massive public outcry and congressional inquiry by the House Oversight Committee.

When asked later if she was sorry for what happened, CEO Heather Bresch replied, "I wasn't going to be apologetic for operating in the system that existed." (As an aside, accountability is when we take responsibility for our own actions, not when we blame our actions on the system.)

The ethical fading was so complete at Mylan that Bresch didn't seem to perceive that she or her company had done anything wrong. Indeed, Bresch mind-bogglingly argued that the EpiPen scandal was a good thing because it brought attention to abuses in the health-care system and served as a catalyst for change. Of course, if Mylan had a culture that placed ethics above earnings and believed its primary responsibility was to its Just Cause—rather than itself or its shareholders—then the company could have used their might in the market to become a champion for change much sooner and with a lot less fuss. Acting unethically, getting caught with your hand in the cookie jar, refusing to accept responsibility for your behavior and then pointing to systemic abuses that made you do those things does not make you Joan of Arc.

Incidentally, two years after the EpiPen pricing scandal, Mylan settled with the U.S. Justice Department for

$465 million for overcharging the government for EpiPens it misclassified as generic rather than branded. As acting U.S. Attorney William D. Weinreb explained, "Mylan misclassified its brand name drug, EpiPen, to profit at the expense of the Medicaid program. . . . Taxpayers rightly expect companies like Mylan that receive payments from taxpayer-funded programs to scrupulously follow the rules." Perhaps Mylan suffers from a severe allergy to acting ethically.

But it can't just be a flawed incentive structure that drives good people to do bad things. If that's all it was, we would expect the people who engage in such behaviors to be consumed by guilt and struggle to sleep at night. By all evidence, though, they seem completely relaxed about the choices they make—and in Bresch's case, defensive and unapologetic. According to social scientists who study the phenomenon of ethical fading, those who commit such violations of trust aren't evil, but they do suffer from self-deception.

Self-Deception

We humans have all sorts of clever ways to rationalize our behavior and deceive ourselves into thinking that the ethically questionable decisions we make are fair and justified, even though a reasonable person would view our actions as quite the opposite. Ann Tenbrunsel, professor of business ethics at the University of Notre Dame, and David Messick, professor emeritus of the Management & Organizations Department at Northwestern University's Kellogg School of Management, are among those who have studied self-deception as a mechanism of ethical fading in organizations. In their work, they identify several uncomfortably simple and common ways that we, as individuals and groups, are able to engage in unethical behavior without perceiving it as unethical.

One of the ways we are able to deceive ourselves comes from the words we use. The use of euphemisms, to be exact. Euphemisms allow us to disassociate ourselves from the impact of decisions or actions we might otherwise find distasteful or hard to live with. Politicians were aware that Americans find torture to be inhumane and inconsistent with our values. So "enhanced interrogation" became the way for them to protect our homeland after September 11 without feeling bad about it.

We do the same thing in business. It is common practice in the working world to choose language that softens or obscures the impact of our behavior. We talk about managing "externalities" instead of talking plainly about "the harm our manufacturing practices cause to the people who work in our factories and to the environment." "Gamification to enhance the user experience" is easier to swallow than "we found a way to get people addicted to our product to boost our results." Human beings become "data points," and "data mining" is a more palpable way of saying that we are tracking people's every click, trip and personal habit. We "reduce head counts," and the online ticket broker charges us a "convenience fee" instead of calling it what it is, a surcharge.

The words we choose can help us distance ourselves from any sense of responsibility. They can, however, help us act more ethically too. Imagine if we actually started calling things what they are within our organizations. If we did, perhaps we would take the time to find more creative, and indeed more ethical, ways of achieving our goals. And in so doing, actually strengthen our cultures in the process. But more on that later.

Another kind of self-deception that contributes to ethical fading is when we remove ourselves from the chain of causation or, as the CEO of Mylan did, blame "the system" for our own transgressions. Sometimes we can take ourselves so far out of the chain of causation that we actually lay all the responsibility for how our products affect a consumer on the consumer. Though it's a legitimate legal concept, caveat emptor, or "buyer beware," is often cited by companies to disassociate themselves from the impact of their decisions. "If they don't like it," the thinking goes, "then they don't have to buy it." This is the oft-invoked response we hear from executives when questioned about their responsibility for the negative effects of their products. Though consumer choice is absolutely a factor, this cannot and does not completely remove an organization from the chain of causation. Yes, the smoker is responsible for the damage they do to their health from smoking, but the cigarette companies are still involved in the chain.

Fulfilling one's legal responsibility does not release a company from their ethical responsibility either. After we click a box to accept their terms and conditions, for example, many companies believe that they are free of responsibility for what happens next. Legally that may be true, but ethically speaking, they are not. Instagram, Snapchat, Facebook and any number of mobile gaming companies, for example, cannot deny their role in making what is increasingly accepted as addictive technology, simply because there is not yet a law against it. Particularly when they knowingly add features such as infinite scroll, "like" buttons, and automatic content play with the intent of keeping us peeled for longer. These companies almost always explain that they add such features or need to collect our personal data in order to "enhance the user experience." Though we may indeed receive some benefit from these decisions, there is also a cost. Weighing those benefits against the harm they may cause or whether they violate our values is what ethics is all about! Nothing is for free.

In a 2019 opinion piece in *The Washington Post*, Mark Zuckerberg, the founder and CEO of Facebook, responded to some of the criticism against his company by asking government for more legislation. "I believe we need a more active role for governments and regulators," he wrote. "By updating the rules for the Internet, we can preserve what's best about it." It's as if he's saying that, because of Milton Friedman's definition of the responsibility of business, Facebook can only be ethical if the laws and "ethical custom" require them to be. It's sad that we have reached a point in some industries, like technology and social media, where we probably do have to legislate ethics. But how did we arrive here in the first place?

Tenbrunsel and Messick identify the proverbial "slippery slope" as another enabler of the kind of self-deception that leads to ethical fading. With each ethical transgression that is tolerated, we pave the road for more and bigger ethical transgressions. Little by little, we change the norms inside a culture of what is acceptable behavior. "If everyone else is doing it, then it must be okay."

When leaders maintain an excessive focus on the finite game, these slippery slopes are often missed or willfully ignored because they are so profitable. In an organization that has adopted an infinite mindset, an unethical idea designed to grow the bottom line is always "a bad idea that we wouldn't touch with a ten-foot pole." In an organization obsessed with the finite game and suffering from fading ethics, that same idea is "fantastic, I can't believe we didn't think of this sooner!" Add an unbalanced reward structure that focuses on performance and ignores trust, and the ethical lapses start to move as if they were sliding down a Slip 'N Slide coated in Teflon covered in baby oil until they reach full-blown ethical fading at the end.

Like the slow boiling of the proverbial frog, Mylan's incremental increase of EpiPen prices was no doubt intended to lessen the shock (or increase the acceptance) of a huge, sudden price increase on consumers. However, it also reveals ethical fading at work. By increasing the price over time (even over a short time) they saw their metrics soar. As the numbers went up, many probably started to imagine what they would spend their bonuses on. Focused on the massive upside they would personally gain, Mylan's executives were able to get ethically comfortable with their decisions. And so they increased the rate of the price increases to hit or beat their goals even quicker. It's as if they were acting like addicts who couldn't wait patiently to get their next fix.

Mylan and Wells Fargo are extreme examples of ethical fading. And such extreme examples are helpful for us to see the mechanics of ethical fading at work. But don't be fooled . . . and don't get comfortable. Just because there is no fraud or scandal doesn't mean we don't have a problem. In fact, if we look closely, we begin to see signs of ethical fading in lots of businesses. Tricks of accounting to reduce a company's tax burden, for example. Or offering a rebate on a product and purposely making customers perform so many steps—cut out the barcode from the box, fill out the form, attach the receipt, mail it in—that the majority of people, as the company knows full well, won't bother doing it, is another. Or food and beverage companies exaggerating the health benefits of a product, attempting to hide some of the unhealthy ingredients or tinkering with the portion size on a package to make it look like their product has less sugar or fewer calories than it actually does. None of it is illegal. All of it is a little uncomfortable. And the more we all allow such decisions to be made, the more such behavior becomes "normal" or the "industry standard."

Remember, ethical fading is about self-delusion. Anyone, regardless of their personal moral compass, can succumb to it. The leaders we point out and vilify for running their businesses unethically and then accepting a handsome reward for doing so don't think they've done anything wrong. And if you don't think you are doing anything wrong, what incentive do you have to do things differently? In a case like Mylan or Wells Fargo, it took a public scandal to expose the problem. But a spotlight doesn't fix the problem. In most of our organizations, there won't be a crisis like those to help see some of the ugly truths. And as long as ethical fading goes unchecked, the odds are high that, eventually, something is going to break. And the cost, not only to our companies, but also to our people, our customers and our investors will be far greater than any cost we would bear to fix things now.

On taking over as CEO at Wells Fargo, Tim Sloan admitted that management "recognized too late the full scope and seriousness of the problems" and vowed that such a situation "will never be allowed to occur again." Such promises are easily made. Not so easily kept. Ethical fading can be extremely difficult to reverse. Almost impossible if the leaders trying to change the culture remain finite minded in their approaches. Because what do finite-minded leaders do when they set out to change a culture that suffers ethical fading? You guessed it. They apply a finite solution. (Hint: It doesn't work.)

When Structure Replaces Leadership

I used to work for a large advertising agency. After my first year at the company, leadership decided to implement time sheets. Unlike a law firm, where a lawyer may be billing their clients for the actual number of hours of work, this was a way for the company to keep track of . . . actually, no one really had any idea of the utility of the time sheets. It was just something we were told to do.

I managed to get away with not filling out mine for months (if they were tracking how I spent my time, I saw no point in telling the company I worked 100 percent on the one client to which I was assigned). Of course, I got in trouble for not turning in my time sheets. And so, from then on, at the end of every month, I sat down with all my time sheets and filled them out in one go—in at 9:30 A.M., out at 5:30 P.M. In reality, I often came in earlier and left later. But who cares. I recall taking my time sheets to my boss for his signature. He looked them over and commented sarcastically, "You're certainly a very consistent worker, aren't you?" And then he signed them.

I have to believe that the time sheets were implemented because something went wrong in accounting. Perhaps a client was overbilled for work done and demanded that the agency prove that the senior people who were promised to spend time on their account actually were the ones who spent time on the account . . . or something like that. In order to correct the issue in accounting, a new process was implemented across the company. This kind of solution is what Dr. Leonard Wong calls "Lazy Leadership."

When problems arise, performance lags, mistakes are made or unethical decisions are uncovered, Lazy Leadership chooses to put their efforts into building processes to fix the problems rather than building support for their people. After all, process is objective and reliable. It's easier to trust a process than to trust people. Or so we think. In reality, "process will always tell us what we want to hear," Dr. Wong points out. "[Process] gives us a green light," he continues, "but it may not be telling us the truth." When leaders use process to replace judgment, the conditions for ethical fading persist . . . even in cultures that hold themselves to higher moral and ethical standards.

Soldiers, for example, believe they hold themselves to a higher standard of honesty and integrity than the general public. And the general public thinks so too. However, in their paper "Lying to Ourselves: Dishonesty in the Army Profession," Dr. Wong and his research partner Dr. Stephen Gerras, both retired army officers who now work at U.S Army War College, discovered systemic ethical fading as a result of excessive process, procedure or demands placed on soldiers. Some of the things leadership was asking of their soldiers weren't unreasonable—they were impossible. Soldiers were required, for example, to complete more days of training than were available in the calendar.

As in the corporate world, pressure to complete tasks comes from the top down in the Army. However, there is also a huge amount of pressure that comes from the bottom up. In an effort to stand out, officers want to appear as if they can do everything and do everything well. A failure to complete requirements could sully a commander's image, earn reprimands and affect promotions. Submitting a false report of compliance helps keep the system running smoothly and keeps their careers on track. And because the punishment for being honest is sometimes greater than for lying, soldiers are put in a position in which they feel they have to lie or cheat in order to meet the requirements placed upon them. It's a Catch-22.

The result is that it has become commonplace for soldiers to find creative ways to complete their requirements while feeling that their high moral standards remain uncompromised. One example Wong and Gerras give involves the last-minute training requirements units had to complete before deploying to Afghanistan or Iraq. Soldiers had to insert their ID cards into a computer to authenticate their identity in order to complete the computer-based training. One officer admitted that he would collect all the ID cards of his nine-man squad, then pick the smartest guy in the group to complete the training nine times so that everyone could get a certificate.

Rather than seeing their actions as cheating or lying, many soldiers saw it simply as "checking the boxes," "part of the bureaucratic process" or just doing what "leadership wanted them to do." Some didn't see their actions as unethical at all because they viewed the demands as so trivial that they existed outside of any standard of integrity or honesty, like me and my time sheets. It's like telling someone we have to cancel plans because of a "family issue" when in reality there is no family issue; we just want to get out of the plans without hurting someone's feelings. And though we told a lie, because it's just a little "harmless" white lie, we still believe ourselves to be honest.

When these seemingly minor transgressions become pervasive in a culture, however, it is a sign of ethical fading. Remember, the very definition of ethical fading is engaging in unethical behavior while believing that we are still acting in line with our own moral or ethical code. As in the corporate world, if any of the unethical acts that the soldiers committed were to lead to more severe consequences that would cause public outrage, it is likely that the soldiers would indeed be punished (and the rest of the Army subjected to additional online training to prevent anything like that from happening again, of course).

There's a great irony in all this. When we apply finite-minded solutions to address an ethical fading problem that finite-minded thinking created, we get more ethical fading. When we use process and structure to fix cultural problems what we often get is more lying and cheating. Little lies become bigger lies. And the behavior becomes normalized.

Lazy Leadership is not a euphemism for bad leaders or bad people. Just like a person who chooses not to exercise is not a bad person. Decisions made by Lazy Leadership can often be very well intended. In the case of the Army, or any large organization for that matter, leadership may genuinely believe all the extra demands and requirements they place on soldiers are helpful. But because senior leaders are rarely subjected to those extra demands themselves, they may be oblivious to the problems their "solutions" cause. However, if they were aware of or also subjected to the hypocrisy, dysfunction or excessive bureaucracy, then like my boss at the agency, they too could become complicit in the charade. When that happens, those leaders are likely also engaging in rationalization and self-deception. And the slope grows slipperier.

If ethical fading can happen in places where integrity is taken really seriously, like the military, then it can happen anywhere. And it does. I cannot stress enough how common ethical fading is in our companies and institutions. However, more structure is not the antidote to ethical fading. Process is great for managing a supply chain. Procedure helps improve manufacturing efficiencies. Ethical fading, however, is a people problem. And counterintuitive though it may seem, we need people—not paperwork, not training, not certifications—to fix people problems.

The best antidote—and inoculation—against ethical fading is an infinite mindset. Leaders who give their people a Just Cause to advance and give them an opportunity to work with a Trusting Team to advance it will build a culture in which their people can work toward the short-term goals while also considering the morality, ethics and wider impact of the decisions they make to meet those goals. Not because they are told to. Not because there is a checklist that requires it. Not because they took the company's online course on "acting ethically." They did so because it's the natural thing to do. We act ethically because we don't want to do anything that would do damage to the advancement of the Just Cause. When we feel a part of a Trusting Team, we don't want to let down our teammates. We feel accountable to our team and the reputation of the organization, not just to ourselves and our personal ambitions. When we feel part of a group that cares about us, we want to do right by that group and make our leaders proud. Our standards naturally rise.

As social animals, we respond to the environments we're in. Put a good person in an environment that suffers ethical fading, and that person becomes susceptible to ethical lapses themselves. Likewise, take a person, even one who may have acted unethically in the past, put them in a stronger, more values-based culture, and that same person will also act in accordance with the standards and norms of that environment. As I've said before, leaders are not, by definition, responsible for the results. Leaders are responsible for the people who are responsible for the results. It's a job that requires constant attention because when little things compound, things eventually break.

Infinite-minded leaders accept that creating a culture that is more resistant to ethical fading requires patience and hard work. It requires devotion to a Cause, a bias for will before resources and the ability to nurture Trusting Teams. It may take longer than a quarter or a year (depending on the size of the company) to feel the impact of the investment. And once the ethical standards are established (or reestablished), they must be guarded vigilantly. If ethical fading is powered by self-deception, maintaining ethical behavior demands complete honesty and constant self-assessment. Ethical lapses happen and are part of being human. Ethical fading, however, is not a part of being human. Ethical fading is a failure of leadership and is a controllable element in a corporate culture. Which means the opposite is also true. Cultures that are ethically strong are also a result of the culture the leaders build.

When Acting Ethically Is the Standard

On November 25, 2011, outdoor clothing company Patagonia took out a full-page ad in *The New York Times* with the headline: "Don't Buy This Jacket." Though some cynics saw the headline as a publicity stunt by a high-priced brand that many people can't afford, it is in the details of the ad that we can find clues about the kind of culture Patagonia has and that inspired such an ad in the first place.

In the body copy of the ad, Patagonia did something most other companies would consider unthinkable. They explained, in plain language, the environmental cost of making their product, in this case the bestselling R2 Fleece. The copy read:

> To make [this jacket] required 135 liters water, enough to meet the daily needs (three glasses a day) of 45 people. Its journey from its origin as 60% recycled polyester to our Reno warehouse generated nearly 20 pounds of carbon dioxide, 24 times the weight of the finished product. This jacket left behind, on its way to Reno, two-thirds its weight in waste.

"There is much to be done and plenty for us all to do," the ad concludes. "Don't buy what you don't need. Think twice before you buy anything. . . . Join us . . . to reimagine a world where we take only what nature can replace."

"We did it out of guilt," says Patagonia founder Yvon Chouinard. "We all know we have to consume less." While other companies might use euphemisms to distance themselves from or cloud the impact of their actions, Patagonia takes full ownership of its role in the chain of causation and offers no exceptions or excuses that might lead the way down a slippery slope. They are brutally honest with themselves and the public about how their actions impact the world, for better or for worse. They know that if they want to survive and thrive in the Infinite Game, they have to be this honest. They don't portray themselves as victims of the system but rather a part of it . . . and they are doing what they can to change it. It's hard to even imagine Mylan taking out an ad in *The New York Times* explaining that they knew they were taking advantage of people with life-threatening allergies by raising the price of EpiPens by 500 percent, claiming they did so to highlight the unethical and legal abuses in the pharmaceutical industry.

The postmortem after any scandal or case of ethical fading nearly always reveals a failure of leadership. Companies with cultures like Mylan and Wells Fargo are almost destined to suffer some sort of ethical fading. With the words of Milton Friedman by their side, their leaders think they are there to drive results, and their incentive structures reinforce that belief. As a result, they prioritize near-term financial results above any sense of Cause (if they even have one). Operating with a bias toward resources before will, the leaders willingly adjust their cultures to meet their priorities. At Patagonia, like any other infinite-minded organization, they turn to their Just Cause to help set their priorities and the behavior follows accordingly. It's not just about how much money they can make this year. "We plan to be here in the next one hundred years, so we think about long-term results," says Dean Carter, vice president of Human Resources and Shared Services at the company. Operating with an infinite mindset, Patagonia's intention is not to win or beat anyone else in their market. Rather, Patagonia is driven by a vision of the future in which they make high-quality products while causing the least harm and "use business to inspire and implement solutions to the environmental crisis."

Patagonia is by no means a perfect company. They make mistakes and individuals within the company still suffer ethical lapses. Patagonia recognizes this and understands that its pursuit of the Just Cause is a journey of constant self-improvement. At too many companies, the term "constant improvement" often means improving process and enhancing efficiency. At Patagonia and other infinite-minded companies, where the currencies of will and resources are both on the radar, constant improvement refers to every facet of their organization, including their culture and the standards by which their culture operates. This is what helps them maintain a culture of high ethical standards. Patagonia is not driven to be the best, they are driven to be better.

Even if the headline fell flat to some, the "Don't Buy This" campaign was not a one-off gimmick. It was typical of Patagonia's relentless effort to hold themselves accountable and constantly improve. As their website says:

> Patagonia is a growing business—and we want to be
> in business a good long time. The test of our sincerity
> (or our hypocrisy) will be if everything we sell is use-
> ful, multifunctional where possible, long lasting, beau-
> tiful but not in thrall to fashion. We're not yet entirely
> there.

The copy goes on to admit that not all of the company's products meet these criteria, but then they go on to introduce their Common Threads Initiative, a program they hope will help advance them toward their goals. The initiative includes a commitment to make high-quality clothes that will last a long time, so they don't have to be routinely replaced (which reduces waste); a promise to repair their products for free, so that people don't throw them out (which reduces waste); a partnership with eBay, so that people can "reuse," buy and sell secondhand products (which reduces waste); and when a product finally does come to the end of its life, Patagonia will take it off your hands to recycle it rather than have us throw it in the garbage (which reduces waste).

While some companies go out of their way to find loopholes they can exploit to enhance their performance, Patagonia goes out of its way to close loopholes that enhance their values and beliefs. For the past decade, for example, the company has been working with the NGO Verité to uncover and correct labor abuses within its first-tier supply chain, in the factories that produce their goods. As a result of separate internal audits in 2011, the company found that despite their efforts to create a socially responsible supply chain, there were still a number of violations, including multiple cases of human trafficking and exploitation, at the second-tier level, the factories that turn raw materials into fabrics and other parts needed for production. It is remarkable that Patagonia was even looking at, let alone trying to improve, conditions in its second-tier suppliers.

"Even the Fair Labor Association (FLA), which conducts spot audits of factories abroad and helps companies improve their corporate-responsibility programs," wrote Gillian White for an article in *The Atlantic*, "only requires that affiliated brands audit, monitor, and report on their first-tier suppliers— a level at which issues of human trafficking are easier to spot and respond to." Rooting out forced labor is a difficult and complex undertaking that requires a major commitment of time and investment of resources. Most companies wait to tackle it only when they are forced to, either out of embarrassment or legal trouble. Patagonia, of its own accord, has made the commitment and the investment, knowing that it may never completely solve the problem. But it will damn well keep trying. Which is the whole point of constant improvement and ethical action. Indeed, it is the very standard of an effective Just Cause—that we may never reach the ideal we imagine but we will die trying. This gives purpose and meaning to the work we do at the companies we work for and inspires us to keep fighting the good fight.

Finite-minded companies might worry that this kind of approach may cost too much, hurt profits, lose customers or ruin their reputation (few companies want to proactively admit they do anything wrong these days). Patagonia is not worried about those things and they're not afraid to get out ahead of the crowd and take big risks. Of course the company has a huge advantage that it freely admits. It's a private company. "The pressure of a public company to drive profit on a quarterly basis for people who only have a financial vested interest in the outcome of the company is significant," Dean Carter reminds us. "So it does help to be private when our vested interest is certainly to make a bigger impact."

Though Patagonia is a certified B Corp—a company that practices "stakeholder capitalism"—it is not a charity. It is a for-profit organization that wants to make more money this year than they made last year. However, they also recognize that making money is not the reason they exist. Like all good infinite-minded companies, they see money as the fuel they need to continue to pursue their Just Cause. To certify as a B Corp, companies are required to identify their most deeply held social and environmental values, then abide by them, honoring their responsibilities to their employees, customers, suppliers and communities—as well as to the financial health of their investors. Patagonia knows that the more successful their business, the more they can uphold that standard and the greater the positive impact they can have in the world. They know that in the long term, if they keep their eye on the Just Cause and continue to guard against ethical fading, they will attract and lead those who share their vision and values and they will thrive as a result.

Ethical decisions are not based on what's best for the short-term. They are based on the "right thing to do." Whereas short-termism at the expense of ethics slowly weakens a company, "doing the right" thing slowly strengthens it. Patagonia's pattern of trying its hardest to do the right thing and put people and planet before profits has earned the company fierce loyalty from employees and customers alike. This, combined with the good will and trust they have built in the market, has helped them become one of the most successful, innovative and profitable companies in their category. The company has experienced a quadrupling of revenue over the past decade, with profits tripling. In the words of Patagonia's CEO Rose Marcario, "Doing good work for the planet creates new markets and makes [us] more money." (Notice the order in which she presents her priorities).

"As far out on the horizon line as we can see right now," says Rick Ridgeway, Patagonia's VP of Environmental Affairs, "we're continuing to produce products that allow people to live a more responsible life with the apparel that they choose. As long as there's a lot of other people out there that don't do that . . . then we *should* be growing." Still, Ridgeway acknowledges, "There is a point out there where our own growth is going to likely create more problems than it does solutions." It remains to be seen how Patagonia will deal with that point if and when they arrive at it. But the very fact that they are thinking about it, and talking about it (publicly no less), is yet another sign of their ethical strength.

As a result of leading with an infinite mindset, Patagonia has not only created a company more resistant to ethical fading, but has also set the bar for what acting ethically can look like in business. And that's by design. "If we can show the business community that we're successful," says COO

Doug Freeman, "we think we're holding ourselves as a great example for how business can be done differently." Patagonia acts in a way that is not just good for them, it's good for the game . . . and it's working. Other companies now follow their lead.

Chapter 9

WORTHY RIVAL

Whenever I heard his name, it made me uncomfortable. If I heard someone sing his praises, a wave of envy washed over me. I know him to be a good person and a nice guy. I respect his work a great deal and he has always been nice to me when we've met in professional settings. We do the same kind of work—write books and give talks about our views of the world. Though there are many others who do work similar to his and mine, for some reason I was obsessed with *him*. I wanted to outdo *him*. I would regularly check the online rankings to see how my books were selling and compare them to his. Not anyone else's. Just his. If mine were ranked higher, I would smile a gloaty smile and feel superior. If his were ranked higher, I would scowl and feel annoyed. He was my main competitor and I wanted to win.

Then something happened.

We were invited to share a stage at the same event. Though we had spoken at the same events before, this was the first time we would actually be on the stage at the same time. In the past I would speak on day one of a conference, for example, and he on day two. This time, however, we would be on stage at the same time, sitting side by side for a joint interview. The interviewer thought it would be "fun" if we introduced each other. I went first.

I looked at him, I looked at the audience, I looked back at him and I said, "You make me unbelievably insecure because all of your strengths are all my weaknesses. You can do so well the things that I really struggle to do." The audience laughed. He looked at me and responded, "The insecurity is mutual." He went on to identify some of my strengths as areas in which he wished he could improve.

In an instant I understood the reason why I felt so competitive with him. The way I saw him had nothing to do with him. It had to do with me. When his name came up, it reminded of me of the areas in which I grappled. Instead of investing my energy on improving myself—overcoming my weaknesses or building on my strengths—it was easier to focus on beating him. That's how competition works, right? It's a drive to win. The problem was, all the metrics of who was ahead and who was behind were arbitrary and I set the standards for comparison. Plus there was no finish line, so I was attempting to compete in an unwinnable race. I had made a classic finite-mindset blunder. The truth is, even though we do similar things, he isn't my competitor, he is my rival. My very Worthy Rival.

To anyone who has spent time watching or playing games and sports, the notion of a finite competition where one player or one side beats the other to earn a title or prize is familiar. Indeed, to most of us, it is so ingrained in the way we think that we automatically adopt an "us" against "them" attitude whenever there are other players in the field, regardless of the nature of the game. If we are a player in an infinite game, however, we have to stop thinking of other players as competitors to be beaten and start thinking of them as Worthy Rivals who can help us become better players.

A Worthy Rival is another player in the game worthy of comparison. Worthy Rivals may be players in our industry or outside our industry. They may be our sworn enemies, our sometimes collaborators or colleagues. It doesn't even matter whether they are playing with a finite or an infinite mindset, so long as we are playing with an infinite mindset. Regardless of who they are or where we find them, the main point is that they do something (or many things) as well as or better than us. They may make a superior product, command greater loyalty, are better leaders or act with a clearer sense of purpose than we do. We don't need to admire everything about them, agree with them or even like them. We simply acknowledge that they have strengths and abilities from which we could learn a thing or two.

We get to choose our own Worthy Rivals and we would be wise to select them strategically. There is no value in picking other players whom we constantly outflank simply to make ourselves feel superior. That has little to no value to our own growth. They don't have to be the biggest players or any of the incumbents. We choose them to be our Worthy Rivals because there is something about them that reveals to us our weaknesses and pushes us to constantly improve . . . which is essential if we want to be strong enough to stay in the game.

From the mid-1970s into the 1980s, Chris Evert Lloyd and Martina Navratilova were two of the dominant players in women's tennis. Though they were competitors when they met on the court, each driven to win, it was the respect they had for each other that helped both of them become better tennis players. "I appreciate what she did for me as a rival, to lift my game," Lloyd said once, speaking fondly of Navratilova. "And I think she appreciated what I did for her." It was because of Navratilova, for example, that Evert had to change the way she played. She could no longer rely on spending time on the baseline. She had to learn to become a more aggressive player. This is what a Worthy Rival does for us. They push us in a way that few others can. Not even our coach. And in the case of Evert and Navratilova, it elevated their own games and the game of tennis.

The impact of this subtle mind shift can be profound in how we make decisions and prioritize resources. Traditional competition forces us to take on an attitude of winning. A Worthy Rival inspires us to take on an attitude of improvement. The former focuses our attention on the outcome, the latter focuses our attention on process. That simple shift in perspective immediately changes how we see our own businesses. It is the focus on process and constant improvement that helps reveal new skills and boosts resilience. An excessive focus on beating our competition not only gets exhausting over time, it can actually stifle innovation.

Another reason to adjust our perspective toward seeing strong players in our field as Worthy Rivals is it helps keep us honest. It's like a runner who is so obsessed with winning, they forget the rules, ethics or why they started running in the first place. They may spend time and energy to undermine someone who is running faster than they are and resort to tripping their competitor. Or perhaps they will take performance-enhancing drugs to give them a secret edge. Both tactics will absolutely increase the chances they will win the race, but such strategies will leave them ill equipped for success beyond those races. And eventually those strategies run dry and they are still left a slow runner. When we view the other players as Worthy Rivals it removes the pressure of being in a win-at-any-cost struggle and so by default we feel less need to act unethically or illegally. Upholding the values by which we operate becomes more important than the score, which actually motivates us to be more honest (organizations or politicians who choose to do the right thing rather than what helps them get ahead are good examples).

As for my Worthy Rival, when I thought of Adam Grant as a competitor, it didn't help me. Rather, it fed my finite mindset. I was more concerned with comparing arbitrary ratings than I was with advancing my own Cause. I devoted too much time and energy to worrying about what he was doing rather than focusing that energy on how I could be better at what I do.

Since that day when I learned to shift my mindset, I no longer compare my book rankings to Adam's (or anyone else's, for that matter). My mindset has shifted away from channeling my feelings of insecurity against him to partnering with him to advance our common cause. We have become dear friends (he kindly gave this book a proofread and helped make it better) and I feel genuine happiness when I hear his name or see that he is doing well. I want his ideas to spread. In fact, everyone reading this book should also read *Give and Take* and *Originals*; they are both essential reading in and out of the business world. (Fun fact: In an infinite game, we can both succeed. Turns out people can actually buy more than one book.) An infinite mindset embraces abundance whereas a finite mindset operates with a scarcity mentality. In the Infinite Game we accept that "being the best" is a fool's errand and that multiple players can do well at the same time.

Worthy Rivals Can Help Us
Get Better at *What* We Do

When Alan Mulally left the airplane manufacturer Boeing Commercial to become the CEO of the ailing Ford Motor Company in 2006, it would be the start of a journey that would result in one of the greatest turnarounds in automotive history. After the formal press conference to announce his new job at Ford, Mulally fielded some questions. One reporter asked what kind of car he drove. "A Lexus," Mulally replied. "It's the finest car in the world." The new CEO of Ford just admitted that the car made by Toyota that he drove was better than anything Ford made! To some it was sacrilege. But to Mulally, a man who prefers the truth, even when it's uncomfortable, it was an honest assessment.

In the 15 years before Mulally took over, Ford had lost 25 percent market share. Now it was headed toward bankruptcy. Indeed, Mulally needed a turnaround strategy, but first he wanted to learn as much as he could about the company. He wanted to understand Ford's health beyond the balance sheet. One of the things he learned was that consumers were disenchanted with the brand. Ford cars (at least in the United States) had a reputation for being unexciting, unreliable gas guzzlers. Perhaps *this* was part of the reason people weren't buying Fords like they used to.

Historically, Detroit's car companies, including Ford,

were obsessed with market share as a primary metric for comparison. However, Mulally knew that some of the most profitable car companies in the world were also some of the smallest. He understood quickly that it wasn't in Ford's long-term interest to just grow market share—something that could be accomplished with sales promotions and cost cutting (which was exactly the turnaround plan Ford presented to Mulally when he arrived). That strategy would only work for a few years. "We're not going to chase market share," he said. "We're not going to put out vehicles where demand is not there and then discount and make it even worse." If Ford was to stay in the game, they would have to change the way they played the game. And that meant it had to relearn to make cars that people actually wanted to drive.

One of the first things Mulally did after joining the company was to start driving home in a different model Ford every night. After trying every single car the company made, he asked to drive home a Toyota Camry. The only problem was Ford didn't have one for him to drive. It was common practice for Ford to buy the cars of other manufacturers so that their engineers could take the car to pieces to see how they are made, but there were none available for anyone to actually drive. Think about that for a second. The senior executives of a major car company that was struggling to sell cars had little idea what anyone else's cars were actually like to drive. If car buyers test-drive their options, shouldn't Ford's executives know what they are trying? Mulally had the company buy a whole fleet of cars made by other companies and instructed his senior managers to drive them.

When he called the Lexus the finest car in the world, Mulally wasn't trying to make the people at Ford feel bad. He was offering them a Worthy Rival. He was convinced that in order to save Ford, they would need to be frank about the state of their own products and processes and respectful students of the other players in their industry. Toyota was a company that, as Mulally describes it, "[makes] products that people want . . . with less resources and less time than anybody in the world." They were a benchmark against which Ford could push themselves to improve the quality of their own cars and how they made them. And if they could pull that off, the profits would follow. For Mulally, the reason to study the other car manufacturers wasn't simply to copy them or outsell them, but to learn from them. "I was never trying to beat GM or Chrysler," Mulally says. "We were always focused on the Just Cause and we used our benchmarking against our competition as data insights on where we could continuously improve our operation." Continuously improving their process would help them make better product, which would help them be more effective at advancing Henry Ford's original Just Cause: to provide safe and efficient transportation for everyone, to open the highways to all mankind. Henry Ford's Cause also served as a filter for other decisions. Mulally sold off brands like Jaguar, Land Rover and Volvo, for example. Ford originally bought them so they could compete in as many automotive categories as possible—something Mulally believed distracted Ford from why the company was founded in the first place.

Then came the 2008 stock market crash, which was particularly devastating for the U.S. car industry. Without a government bailout, GM and Chrysler would go bankrupt. Thanks to a nearly $24 billion loan that Mulally had taken out in 2006 to help Ford reinvent itself, combined with the steady improvements the company was making in its operations and products, Ford would be able to weather the downturn without any government assistance. So when Mulally showed up to testify in front of Congress before the bailouts were given, he could have insisted that the government not give money to GM or Chrysler. A CEO who sees the other players as their competitors would have relished watching them go bankrupt, leaving Ford as the only major U.S. car manufacturer to survive. Surely that's winning?

Because Mulally saw the other makers as Worthy Rivals, he actually endorsed the bailout. He knew that keeping those companies around would only serve to help make Ford a better company. He also knew that Ford's rivals were part of a larger ecosystem. If they went bankrupt, so would many of the suppliers. Which could also destroy Ford. So Mulally put together plans to also help many of the auto suppliers weather the downturn. Unfortunately, the leaders of the troubled GM and Chrysler, still operating with a finite mindset, rejected Ford's request to work together for the good of the industry. In contrast, Honda, Toyota and Nissan did work with Ford to help keep major suppliers, on which they also relied, in business. The infinite-minded players understood that the best option for their own survival, and indeed the ultimate goal of an infinite leader, is to keep the game in play.

Worthy Rivals Can Help Us
Get Clearer on *Why* We Do It

By the early 1980s, the computer revolution was in full swing. And for Apple, one of the companies that was leading the computer revolution, the true value of their rival had little to do with product improvement. It was bigger than that. Their Worthy Rival helped them better clarify their Cause and rally their people. The mere existence of their Rival reminded everyone inside and outside the company what they stood for—the reason they went into business in the first place. "They were the navy. We were the pirates."

During the 1970s, IBM had the lion's share of the market in mainframe computers—huge, room-filling machines that offered companies massive computing power. But IBM resisted developing their own "microcomputers," as they used to be called, believing them to have insufficient computing power to meet a business's needs. Personal computers, IBM believed, had no place in the office.

That all changed in 1981. Seeing how well the pioneers of personal computing—Commodore, Tandy and Apple— were doing in getting their products to businesses, IBM changed its tune. Flush with cash, IBM was able to invest massive amounts of money to develop their own personal computer. They paid exorbitant salaries to steal some of the best and brightest engineers in the business from other companies, including Apple. And in just twelve months, IBM introduced its "PC" to the world.

Apple had the biggest market share in personal computers before IBM showed up. Which meant that they had the most to lose when IBM entered the market. Whereas a finite-minded player would likely panic at such news, an infinite-minded player, like Apple, did the exact opposite. In August 1981, in the same month IBM's PC first went on sale, Apple ran a full-page ad in *The Wall Street Journal* with the headline: "Welcome, IBM. Seriously." The rest of the ad tells us everything we need to know about how Apple viewed this new player—not as a competitor, but as a Worthy Rival.

"Welcome to the most exciting and important marketplace since the computer revolution began 35 years ago," read the opening sentence of Apple's ad. "Putting real computer power in the hands of the individual is already improving the way people work, think, learn, communicate and spend their leisure hours," the ad continued. "Over the next decade, the growth of the personal computer will continue in logarithmic leaps. We look forward to responsible competition in the massive effort to distribute this American technology to the world. And we appreciate the magnitude of your commitment. Because what we are doing is increasing social capital by enhancing individual productivity." Apple signed the letter to their new Rival with the words: "Welcome to the task." Apple was trying to advance a Just Cause, and IBM was going to help them.

IBM accepted the challenge. And because of their dominance in the business world, IBM was able to leverage those relationships to sell their new personal computers into large companies. This made Big Blue, as IBM was affectionately called, the safe and obvious choice for any procurement manager who was responsible for buying PCs for their company. "No one ever got fired for buying IBM," the saying went. To further grow their business, IBM allowed other computer makers to "clone" or use their operating system in their products. Apple refused to follow suit. If someone wanted Apple's operating system, they had to buy an Apple. Unable to clone Apple's OS and because it was expensive to develop another operating system for the mass market, most other computer makers licensed IBM's operating system to produce IBM-compatible products. And with that, the PC became the industry standard in the business world and beyond.

IBM helped Apple turn the personal computer into a necessity on every desk and a basic household appliance in every home. But IBM did much more than that for Apple. Apple used IBM as a foil to help tell the story of what they stood for in a way that was clearer and more compelling. Just Causes exist in our imaginations, but companies and products are real. And for a person or a company with a clear sense of Cause, that individual or organization itself can become the tangible symbol of their intangible vision. It's easier for us to follow a real company or a leader than an abstract idea. And it's easier to form a compelling narrative for our Just Cause when we can point to a tangible representation of the alternative.

"They were the navy, predictable, sold to corporations," is how John Couch, one of Apple's early employees, described IBM. "We wanted to be the pirates that empowered individuals to be creative." Like Republicans and Democrats, like the Soviet Union and the United States, IBM and Apple stood as symbols of alternative ideologies looking for followers. IBM represented business, stability and consistency. Apple stood for individuality, creativity and thinking differently. By playing up the contrasts to the public, Apple moved from being a leader in the personal computing revolution to being a leader of like-minded revolutionaries.

Based on the standard metrics against which we measure the quality of a computer—price, speed and memory, for example—PCs and Apples were basically equal. In fact, the IBM clones were often quite a bit cheaper. Where competitors almost always only compare the features and benefits of their products, Apple chose to engage with IBM on a level higher than that. Competitors compete for customers. Rivals look for followers. To Apple's followers, IBM was the past and Apple was the future. And to IBM devotees, Apple was a toy for creative types and IBMs were for serious people doing serious work. This was bigger than products and features. This was now a game of religion.

The manner in which Apple responded to IBM entering the PC market was the total opposite of what normally happens. When a new company joins an industry with such force, it often spooks the incumbents. They frequently lose sight of their vision and start focusing on competing with the new player based on product comparisons and other standard metrics. Which means, if they weren't already playing with a finite mindset before, the choice to view the new entrant as a competitor rather than a Worthy Rival will drag them into the finite quagmire before too long. This is exactly what happened to the Canadian cell phone maker BlackBerry.

Over a quarter of a century after IBM stormed into Apple's market, Apple did the same thing to BlackBerry. Except unlike Apple's choice to view IBM as a Worthy Rival that could help them better clarify what they stood for, BlackBerry chose to see Apple as a competitor to be beaten. And they paid a hefty price for that finite-minded decision.

Before the iPhone, BlackBerry was the second largest cell phone operating system in the world. Their high performing, highly durable and very reliable products made them the must-have option in government and in many companies. They owned the business market. Even after Apple introduced their iPhone in 2007, BlackBerry's momentum continued to carry them to a record high 20 percent share of cell phone sales in 2009. As iPhones became more and more popular, however, BlackBerry panicked. BlackBerry's leaders could have chosen to draw a contrast between their philosophies and Apple's, as Apple had done with IBM decades before. They could have used Apple as a foil to highlight their own vision of the world, one that revolved around the security and reliability needs of business and government. But they didn't. Instead, BlackBerry responded to the iPhone's rising popularity by trying to copy it. First, they started offering apps and games for their existing devices, which dramatically slowed their products' performance. Then they abandoned their iconic, full QWERTY keyboards and introduced touch screen options. They never really worked as well as iPhones and were much less durable than their other models.

Sadly, this is a common scenario. Disruption, remember, is often a symptom of a finite mindset. Leaders playing with a finite mindset often miss the opportunity to use a disruptive event in their industry to clarify their Cause. Instead, they double down on the finite game and simply start copying what the other players are doing with the hope that it will work for them too. And in the case of BlackBerry, it didn't. They abandoned the chance to be leaders for a Cause and opted to become followers of a product. Obsessed with trying to beat Apple, they actually lost sight of their own vision. They forgot why they went into business in the first place. And in short order, BlackBerry went into a steep and steady decline. By 2013 the company had less than 1 percent market share, a nearly 99 percent drop in just four years. Where once they dominated, today BlackBerry is an insignificant player and no company's Worthy Rival.

IBM served as Apple's Worthy Rival for many years. Eventually, as computers became ubiquitous and the market changed, IBM dropped out of the PC game. The loss of their Rival did not mean Apple won, however. They quickly found a new symbol of safe, stable, corporate-mindedness in Microsoft ("I'm a Mac, I'm a PC," for those who remember). Like IBM, as Microsoft's own Cause became fuzzier, it no longer offered the clear ideological contrast to Apple that it once did. So who is Apple's Worthy Rival now?

Perhaps Apple's new Worthy Rivals are Google and Facebook. Google and Facebook now represent the Big Brother of the internet; always watching us, tracking our every move in order to sell our data to companies who want to target their advertising to us (which helps Google and Facebook make more money). This has become an "industry standard." Apple still seems to be fighting for the rights of individuals and challenging the status quo. The company has become an outspoken advocate for individual privacy. Unlike their Rivals, Apple has decided not to sell the data they collect as a means of driving revenue. They have also stood up against the government and denied them access to our private text messages. Even though the world around them has changed, for over 40 years Apple has found Worthy Rivals to help keep them focused on the very cause upon which the company was founded.

Cause Blindness

I have a friend who is so focused on her Cause, it is as if she has forgotten that there are other points of view in the world besides her own. My friend, sadly, has labeled anyone who has a different opinion as wrong, stupid or morally corrupt. My friend suffers from Cause Blindness.

Cause Blindness is when we become so wrapped up in our Cause or so wrapped up in the "wrongness" of the other player's Cause, that we fail to recognize their strengths or our weaknesses. We falsely believe that they are unworthy of comparison simply because we disagree with them, don't like them or find them morally repugnant. We are unable to see where they are in fact effective or better than we are at what we do and that we can actually learn from them.

Cause Blindness blunts humility and exaggerates arrogance, which in turn stunts innovation and reduces the flexibility we need to play the long game. Less able to engage in any kind of honest or productive practice of constant improvement, we end up repeating mistakes or continue to do many things poorly. Plus, hubris increases the chance that any weaknesses our organization may have are left open to exploitation by other players. All of which contributes to the draining of will and resources we need to stay in the game. Whenever I try to show my friend that those players she finds despicable are really good at certain things and she should respect them for that, she mocks me and thinks me a turncoat because I dare pay her competitor a compliment.

As hard as it may be to recognize a player as one of our Worthy Rivals, especially if we find them disagreeable, to do so is the best way to become better players ourselves. "The more I questioned these guys, the more I came to understand that the successful criminals were good profilers," explained John Douglas, retired FBI unit chief and pioneer of criminal profiling. Douglas understood that, as unconscionable as we all find serial killers, for example, the best way to catch one was to acknowledge that they were very good at the exact same thing that the FBI does . . . which meant the FBI had to be better. Having Worthy Rivals—criminals adept at evading the FBI—pushes the FBI to constantly improve their techniques.

Having a rival worthy of comparison does not mean that their cause is moral, ethical or serves the greater good. It just means they excel at certain things and reveal to us where we can make improvements. The very manner in which they play the game can challenge us, inspire us or force us to improve. Who we choose to be our Worthy Rivals is entirely up to us. And it is in the best interest of the Infinite Game to keep our options open.

Don't Confuse Losing Your Worthy Rival with Winning the Game

It was soon after the fall of the Berlin Wall that the United States committed what may have been one of the greatest foreign policy blunders of the 20th century. America declared that it had "won" the Cold War. Except it hadn't. By this point in the book, we all know the mantra: in the Infinite Game, there is no such thing as winning. This is true in business or in global politics. America didn't win the Cold War. The Soviet Union, drained of will and resources, dropped out of the game.

The Cold War met all the standards of an Infinite Game. Unlike finite warfare, where there are agreed-upon conventions for play, easily identifiable sides and a clear definition of when the war will end (e.g., a land grab or some other easily measurable, finite objective). In stark contrast, the Cold War was often played out with proxy players, there were no ground rules and there was certainly no clearly defined objective that would signal to all sides that the war will end. As much as the United States and the West talked about "defeating" the Soviet Union and "winning" the Cold War, short of an all-out nuclear war—which was something neither side wanted—few could imagine or predict exactly what winning looked like. And there was no treaty that ended the Cold War. Instead, both sides kept playing, always trying to improve the manner in which they played, with an unknown sense of where it was all going. So when the Berlin Wall came down in 1989, it was not something either predicted would happen.

Like in business, times change and so do the players. And, like in business, if a big company goes bankrupt, it doesn't mean the game is over or that any company is the winner. The players left standing know that other companies will rise up and new ones will join the industry. When our most important Worthy Rival, the one who pushes us more than any other, drops out of the game, it does not mean that there are others on the bench waiting to immediately rush in to play either. It can take years for a new or different Rival or Rivals to replace them. The advanced player in the Infinite Game understands this and works to remain humble at the loss of a major Rival. Cautious not to let hubris or a finite mindset take hold, they play knowing that it is just a matter of time before new players emerge. Patience is a virtue in infinite play. This was not how America acted.

After the Soviet Union left the game, America suffered a sort of Cause Blindness and believed itself to be unrivaled. And so, it acted accordingly. It acted like a victor. Even if well intentioned, it started to impose its will on the world, unchecked, for about 11 years. It anointed itself the world's police force, sending troops to the former Yugoslavia, for example, and imposing no-fly zones over sovereign nations. Things that would have been much harder, if not impossible, to do if the Soviet Union were still around. Without identifying our Worthy Rivals, strong players start to falsely believe they can control the direction of the game or the other players. But that's impossible. The Infinite Game is like a stock market; companies list and delist but no one can control the market.

Highly successful players with lots of money and many strengths can get away with ignoring their weaknesses for a while. But not forever. Fast-growing companies with strong products, marketing and balance sheets, for example, often neglect to give time and attention to leadership training or to actively nurturing their culture. Things that can come back to haunt them later. Groupon is just one example. Hailed by the business press for their product innovation and rate of growth, the leaders neglected their people. Which, when the growth slowed and other companies matched their product, became their Achilles' heel. Uber is another example. They may have pioneered ridesharing technology, but the company has suffered more because of a neglected culture than any product failing. When Dara Khosrowshahi replaced Travis Kalanick as CEO in 2017, it was done with the express purpose of fixing the company culture.

America would have been well served to look for new Worthy Rivals that may have helped the nation prepare for the next chapter of the Cold War. The nation's leaders could have looked beyond strengths like military and economic might to focus on some of the weaknesses they had been neglecting for so many years. But that's not what happened. Relying on the manner of play it had developed and perfected during the years of Cold War 1.0, America was unable to see the rise of new Rivals that aimed to check its actions and ambitions.

Cold War 2.0

There are three tensions that govern the Cold War—nuclear, ideological and economic. (Not coincidentally, these things overlap with Life, Liberty and the pursuit of Happiness as stated in the Declaration of Independence. For America and all nations, these things are existential. They are the things worth bearing any burden or paying any price to defend). During Cold War 1.0, all three of those tensions were conveniently colocated in a single Rival—the Soviet Union. The two nations each possessed more nuclear weapons by an order of magnitude greater than all other nuclear-armed nations combined. Both nations were ideological exporters looking for customers and allies. America was spreading the gospel of democracy and capitalism and the Soviets were proselytizers of communism. And their economies were the two largest economies in the world from the end of World War II until the fall of the Berlin Wall—the entire length of Cold War 1.0.

Having one primary Worthy Rival has huge advantages. It provides for a single point of focus for strategies to be developed, resources to be allocated and the attentions of internal factions to be pointed. Much was written after the events of September 11, 2001, about the lack of cooperation among America's intelligence services, for example. This wasn't a new development. Those agencies were always territorial and competitive with each other. The difference was, when America knew who its Worthy Rival was, when push came to shove, all the agencies could put aside their internal gripes to come together to face the common threat. Absent the identification of any new Worthy Rivals, the internal fighting among so many of America's institutions continued unchecked. Even Republicans and Democrats used to be able to agree that the Soviet Union represented a greater threat to the United States than each other and could always come together in a clear common cause. That is no longer the case. Absent an identified external Worthy Rival, the two parties now see each other as the existential threat to the nation. All the while, the real threats to America grow ever stronger.

So while America was focusing its energies against itself, it failed to see that the Cold War was still alive and well. Except, unlike during Cold War 1.0, in Cold War 2.0, there is not one Worthy Rival, but many. The nuclear threat posed by the Soviet Union has been replaced by North Korea and others. The Soviet economic rivalry has been replaced by China (which is on course to surpass America's economy). The ideological rivalry that the Soviet Union represented has been replaced by extremists acting under the guise of religion. Plus Russia still continues to test and check America's resolve when possible across all three tensions too.

Like in business, the emergence of new players necessarily changes the way the game must be played. Blockbuster—the sole superpower in the movie rental business—failed to appreciate that a small company like Netflix and an emerging technology like the internet required them to reexamine their entire business model. Big publishers doubled down on old models when Amazon showed up instead of asking how they could update and upgrade their models in the face of a new digital age. And instead of asking themselves, "What do we need to do to change with the times," taxi companies chose to sue the ridesharing companies to protect their business models instead of learning how to adapt and provide a better taxi service. Sears got so big and so rich from sending out paper catalogues for so many decades that they were too slow to adapt to the rise of big-box stores like Walmart and ecommerce. And believing itself without Rival, the behemoth that was Myspace didn't even see Facebook coming. What got us here won't get us there, and knowing who our

Worthy Rivals are is the best way to help us improve and adapt before it's too late.

Without a Worthy Rival we risk losing our humility and our agility. Failure to have a Worthy Rival increases the risk that a once-mighty infinite player, with a strong sense of Cause, will gently slide into becoming just another finite player looking to rack up wins. Where once the organization fought primarily for the good of others, for the good of the Cause, without that Worthy Rival, they are more likely to fight primarily for the good of themselves. And when that happens, when the hubris sets in, the organization will quickly find its weaknesses exposed and too rigid for the kind of flexibility they need to stay in the game.

Chapter 10

EXISTENTIAL FLEXIBILITY

Some thought him mad. He began liquidating his assets and selling off property. He borrowed against his life insurance policy and even licensed the rights to his own name. With the company doing so well, why would he leave now to do something different, something so risky? But in 1952, that's exactly what Walt Disney did. He hadn't gone mad. What he had done was make an Existential Flex.

Walt Disney was accustomed to taking risks and doing new things. As a young artist working in the emerging field of animation, Disney was constantly innovating. He was one of the first to make short films in which real actors would interact with cartoon characters. In 1928, he was the first to make a cartoon with synchronized sound, in the animation classic *Steamboat Willie*. Dissatisfied with just making entertaining shorts designed to make an audience smile, however, Disney set out to make an animated film that was a believable substitute for reality. One that could elicit the full range of human emotions. And in 1937, he released the first-ever feature-length animated film—*Snow White and the Seven Dwarfs. Snow White* was like nothing the world had ever seen before. This evolution of Disney's work wasn't the result of experimentation for the sake of experimentation. Nor was it driven by a desire to get rich or become famous. With each step, Disney was advancing his Just Cause, inviting audiences to leave the stresses and strains of life behind and enter into a more idyllic world of his creation.

The seeds of Disney's Just Cause were planted when he was just a boy. When he was four years old, his father, Elias, moved the family from Chicago to live on a farmstead in rural Marceline, Missouri. The young Walt played outdoors, where there were often animals roaming around, where he was surrounded by extended family and a supportive community. It was, as his older brother Roy later recounted, "just heaven for city kids." But that idyllic childhood didn't last long. Elias Disney's attempt at being a farmer ended in failure, and five years after they arrived in Marceline, the family was forced to move again.

Landing in Kansas City, Elias bought a paper route and the young Walt was put to work to help the family make ends meet. But it didn't help. And as their financial struggles compounded, so did Elias's stress . . . and temper. "It reached a point," Walt Disney recalled, "that to tell the truth with my father got me a licking." Fortunately, back on the farm, Disney had discovered drawing, a hobby that gave him a perfect escape from what he perceived as the hardships of real life. For the rest of his days, Disney would use his art and imagination to offer others a chance to escape their present circumstances too, to take them to a place where they could experience the kind of joy he remembered from his childhood in the little town of Marceline.

To Infinity and Beyond

Walt Disney's ability to transport people to another world turned out to be quite profitable. In addition to the critical and popular acclaim, when *Snow White* came out it grossed over $8 million in its first year alone (an equivalent of over $140 million in today's dollars). With the money and success generated from the film, Walt built a studio in Burbank, California, and a corporate culture that was, as former employee Don Lusk described, "just heaven." To repay the debt accumulated from building the studio and fuel their growth, Roy Disney, CEO of Walt Disney Productions, wanted to take the company public. Walt opposed the idea for fear that shareholders would meddle in the business. Eventually, however, Disney succumbed to the pressure and the company did go public.

As the company grew, it faced a host of new challenges. For starters, the culture of Walt Disney Productions became more stratified. Perks that used to be offered to everyone, for example, were now only offered to more senior people. And as wage gaps increased, so did internal dissent. And for the first time, Disney faced hostile struggles with unions. The breakdown of Disney's studio Utopia, combined with pressure to do more cost-conscious live-action movies and the creative restriction he felt from the bureaucracy, left Disney feeling downright defeated about the future. Walt Disney Productions had become more finite and less visionary and

Disney became convinced that the business could no longer serve as a mechanism to advance his Just Cause. Despite his frustrations, his vision remained as infinite as ever. Which is why Disney decided to make an Existential Flex. So he quit. Fifteen years after the original release of *Snow White*, Walt Disney left to do something new.

Taking all the money he made from selling property, other assets and his shares in Walt Disney Productions, combined with a loan he took out against his life insurance policy, in 1952 Disney formed a new company. He called it WED, after his initials, and set to work on a new project, one that he believed could advance his Cause more than anything that came before—an actual place where people could go to escape the reality of their everyday lives. He was going to build the happiest place on earth. He was going to build Disneyland.

Unlike the often dangerous and dirty amusement parks that existed in the day, which tended to be just collections of random rides, the place Walt wanted to build would be safe and immaculate and have a coherent story that ran throughout the park. There would be no signs of toil or trouble, nothing dark and seedy lurking in the shadows. Here, people would be fully immersed in a perfect illusion. "I think what I want Disneyland to be most of all is a happy place—a place where adults and children can experience together some of the wonder of life, of adventure, and feel better because of it," said Disney. It is a place where "you leave TODAY . . . and enter the World of YESTERDAY and TOMORROW."

Whereas audiences could only watch movies, at Disneyland, they could be in the movies. And unlike a movie, which is finite, the park was something that could keep evolving forever. In true infinite-minded fashion, Disney explained:

"Disneyland will never be finished. It's something we can keep developing and adding to. A motion picture is different. Once it's wrapped up and sent out for processing, we're through with it. If there are things that could be improved, we can't do anything about them anymore. I've always wanted to work on something alive, something that keeps growing. We've got that in Disneyland."

Like so many entrepreneurs, Walt Disney put everything on the line when he started his businesses. Setting out to build Disneyland, however, was perhaps the biggest risk of all because he didn't have to do it. He had a lot more to lose than he had had the first time. This is the plight of the infinite-minded, visionary leader. Once he realized that the company was on a path that could no longer advance his Cause, he was willing to put everything on the line to start over again. He didn't leave because he saw an opportunity to make more money. He didn't leave a failing business. He found a better way to advance his Just Cause and he leapt at it.

The Vision for Flexibility

Existential Flexibility is the capacity to initiate an extreme disruption to a business model or strategic course in order to more effectively advance a Just Cause. It is an infinite-minded player's appreciation for the unpredictable that allows them to make these kinds of changes. Where a finite-minded player fears things that are new or disruptive, the infinite-minded player revels in them. When an infinite-minded leader with a clear sense of Cause looks to the future and sees that the path they are on will significantly restrict their ability to advance their Just Cause, they flex. Or, if that leader discovers a new technology that is more likely to help them advance their Cause going forward than the technology they are currently using, they flex. Without that sense of infinite vision, strategic shifts, even extreme ones, tend to be reactive or opportunistic. Existential Flexibility is always offensive. It is not to be confused with the defensive maneuvering many companies undergo to stay alive in the face of new technology or changing consumer habits. Many newspapers and magazines uprooted their business models when they went digital, for example, not because they found a better way to advance their Cause but because they were forced to make the change in the face of a changing world. Though necessary to stay alive, that kind of change rarely inspires the people inside the organization or reignites their passions. An Existential Flex does.

Many start-ups are fueled more by an entrepreneur's passion for a vision than by resources they have to advance it. An Existential Flex recreates that passion for something new at a time when the company is already enjoying success. When Walt Disney started over again with WED, he brought a group of people from the original company who wanted to go on the new adventure with him as if it were the first time. They were willing to share the risk, they were willing to put in the hours, they were willing to do whatever they had to do to make this new idea successful. They found Disney's enthusiasm infectious and were excited to, once again, do things they never dreamed of. The Flex also rejuvenated Disney's own passion. "Dammit, I love it here!" he said of his new company.

An Existential Flex doesn't happen at the founding of the company, it happens when the company is fully formed and functioning. To all the finite-minded observers, it is existential because the leader is risking the apparent certainty of the

current, profitable path with the uncertainty of a new path—
which could lead to the company's decline or even demise.
To the finite-minded ·player, such a move is not worth the
risk. To infinite-minded players, however, staying on the current path is the bigger risk. They embrace the uncertainty.
Failure to flex, they believe, will significantly restrict their
ability to advance the Cause. They fear staying the course
may even lead to the eventual demise of the organization.

Again, the motivation for an infinite-minded player to
Flex is to advance the Cause, even if it disrupts the existing
business model. To the finite-minded player, the reason not
to Flex is expressly to protect the current business model,
even if it undermines the Cause. And if the company is the
vehicle a leader uses to advance their Cause, then making a
dramatic shift in strategy to keep a company going for a very
long time, in one form or another, is also of paramount importance in the Infinite Game.

Existential Flexibility is bigger than the normal day-to-
day flexibility required to run an organization. And we must
not confuse shiny-object syndrome with Existential Flexibility, either. There is a whole category of frustrated employees
around the world who work for well-meaning, sometimes visionary leaders who, like a cat reacting to a shiny object, want
to chase every good idea they come across with "This is it!
We have to do this to advance the vision!" When an Existential Flex happens, it is clear to all those who believe in the
Cause why it has to happen. And though they may not enjoy
the upheaval and short-term stress such a change may cause,
they all agree it is worth it and want to do it. Shiny-object
syndrome, in contrast, often leaves people flummoxed and
exhausted rather than inspired.

When a visionary leader makes an Existential Flex, to the

outside world it appears that they can predict the future. They can't. They do, however, operate with a clear and fixed vision of a future state that does not yet exist—their Just Cause—and constantly scan for ideas, opportunities or technologies that can help them advance toward that vision. Alan Mulally used his business plan review meetings at Ford to also look at what was happening in companies beyond his traditional competitors. "It's about always keeping an eye on all the things that are going on and learning from that," he explained. Where a more finite-minded leader is also looking for opportunities, their gaze tends to be within their industries, on the balance sheet or toward the horizon. An infinite-minded leader with a Just Cause looks outside their industry and miles beyond the horizon—to a place that requires imagination to see. This was certainly the case when Steve Jobs made an Existential Flex at Apple in the early 1980s.

As I wrote about in the previous chapter, Apple had a very clear sense of Cause. And the seeds of that Cause were sown long before Apple was founded. Growing up in Northern California during the Vietnam War, the company's founders, Steve Jobs and Steve Wozniak, were deeply mistrustful of the establishment. They loved the idea of empowering individuals to stand up to Big Brother. During the computer revolution of the 1970s, the two young entrepreneurs saw the personal computer as the perfect tool for individuals to challenge the status quo. They imagined a time in which, thanks to the personal computer, individuals would have the power to stand up to a corporation, maybe even compete with them.

After the launch of the Apple I and the Apple II, Apple was already a highly successful company. They were working on their next product iteration at the time when, in December 1979, Jobs and a handful of his executives visited Xerox

PARC, Xerox's innovation center in Palo Alto, California. While on the tour, the Apple executives were shown one of the new technologies Xerox had developed, called the "graphical user interface." The graphical user interface allowed people to use a computer without learning a computer language like DOS. Instead, with GUI, users could, for the first time, use a "mouse" to move the "cursor" on the screen to "click" on visual "icons" and "folders" that were sitting on the "desktop." If the vision was to empower individuals, this one innovation would make it possible for even more people to use computers than could before.

After the Apple executives left Xerox PARC, Jobs shared his idea. Apple had to change the course they were on. They had to invest in GUI. One of the executives, attempting to be a voice of reason, spoke up. "We can't," he said. He reminded Jobs that Apple had already invested millions of dollars and countless man-hours on an entirely different direction. Abandoning that work to ostensibly build a new product from scratch would add significant strain on the company. According to Apple folklore, the executive went on to say: "Steve, if we invest in this, we will blow up our own company." To which Jobs replied, "Better we should blow it up than someone else."

A more finite-minded leader would be hard pressed to simply walk away from an established strategic path, especially if it included walking away from any significant time or money that had already been invested or the promise of a performance bonus. Despite the cost and the stress it would put on the company, to Jobs an Existential Flex was Apple's only option. The Just Cause directed his choice, not the cost of the choice. And Apple's employees agreed. The people who loved working at Apple loved that Jobs pushed them to do things that neither they nor anyone else had done before. And with that, they set themselves on a path that in just four years saw the introduction of the Macintosh. A computer operating system that completely revolutionized personal computing. For the first time, the personal computer really was easy enough for just about anyone to use. Microsoft was forced to follow Apple's lead. Nearly four years after the introduction of the Mac, Microsoft released Windows 2.0, the first version of Windows to look and feel like the version many use today. It is software that was designed to make a PC work like a Macintosh.

If You Don't Blow It Up, Someone Else Will

"As convenient as the pencil," said the advertising. "You press the button, we do the rest." That pretty much summed up George Eastman's vision when his company, Eastman Kodak, introduced the first cameras ever to be sold to the general public. This was the late 19th century and photography was almost exclusively performed by professionals and serious hobbyists back then. Regular people just couldn't take their own pictures of their families or vacations. The equipment was bulky and heavy. The chemicals required to treat the photographic plates were highly toxic. It was complicated and expensive. But Eastman was obsessed with simplifying photography for the masses. Though he made early advances in how photographic plates were coated, his real breakthrough came when he invented a way to completely replace the heavy plates with a type of cellulose nitrate plastic called celluloid. Originally used to make the dental plates that hold in dentures, we are more familiar with its modern use: film.

Having brought photography to the masses, Kodak went

on to become one of the biggest companies in the world and George Eastman one of the richest men. And following his death in 1932, the company went on to further advance Eastman's Cause. Always looking for ways to give average folks better ways to capture their own memories, in 1935 Kodak introduced the first commercially successful color film for the masses. This also paved the way for color motion pictures and color home movies. Kodak also invented the slide projector with a round tray that made it easier and more convenient for people to share the pictures of their vacations, weddings and anything else they could force their friends and family to sit down to watch. In the early 1960s, Kodak invented the film cartridge, making photography even simpler and more convenient. Now, people who struggled with or were intimidated by threading film onto a spool in their camera only had to plop the film cartridge into the back of the camera and they were off to the races (literally . . . if that was their thing). And in 1975, the R&D department developed something truly remarkable: the first digital camera. But there was a problem . . .

Though going digital was an obvious next step for the company to advance its Just Cause, the problem was, the invention of digital photography directly challenged the company's business model. Kodak made money on every part of taking pictures. They made the cameras, the film, the flash cubes, the machines that processed the film, the chemicals that were used to develop the film and the paper the pictures were printed on. Everyone knew that the new digital technology would render their current business obsolete. If George Eastman or any other infinite-minded leader were at the helm, this wouldn't be an issue. They would see the new technology as a better way to advance their Cause and they would figure out how to reconfigure their company. Sadly, the Cause had been brushed to the side at the executive levels. Finite thinking now dominated. They were no longer making decisions to advance the Cause, they were making decisions to manage the costs and maximize their near-term financial standing.

Lacking any sense of vision whatsoever, when the executives at Kodak were first shown the digital technology, their initial reaction was that people would never want to look at pictures on a screen. The executives told their engineers that people liked their pictures on paper and there was nothing wrong with paper. Steven Sasson, the young engineer who is credited with inventing the digital camera, tried desperately to get the executives to imagine the future of photography 20 or 30 years ahead. Much to his dismay, his leaders had no interest in advancing the Cause and certainly no stomach for any decision that would upset the status quo, especially when the status quo was working just fine and was quite profitable for them personally. They had no appetite to upset Wall Street or go through what would have been the short-term hell of blowing up their own company in order to advance their Just Cause and remake Kodak into a digital company.

And so, abandoning Eastman's vision, instead of making the Existential Flex they needed to, they instead decided to suppress the new technology for as long as they could to stave off the inevitable. "When you're talking to a bunch of corporate guys about 18 to 20 years in the future, when none of those guys will still be in the company, they don't get too excited about it," said Sasson. "Every digital camera that was sold took away from a film camera and we knew how much money we made on film," Sasson continued. "Of course, the problem is pretty soon you won't be able to sell film—and that was my position." Instead of leading the digital revolution, Kodak's executives chose to close their eyes, put their fingers in their ears and try to convince themselves that everything was gonna be just fine. And I guess it was . . . for a time. But it didn't last. It couldn't last. Finite strategies never do.

Now that the digital genie was out of the bottle, Kodak predicted that it would take about 10 years for other companies to seize on digital photography and make it a thing. And they were right. About 10 years after they first invented the digital camera, Nikon, the Japanese camera company, introduced an SLR camera that gave users the ability to attach an external digital processor (which was made by Kodak because they owned the patents) to the body. But it was Fuji, a much smaller Japanese film company, that, in 1988, exactly 100 years after Eastman introduced the first film camera for the masses, introduced the first fully digital camera to the market. Nikon later partnered with Fuji, and together they continued to innovate and refine the technology. About 10 years after that, Sharp, a Japanese electronics company, introduced the first cellular phone. And 10 years after that, by the mid- to late 2000s, digital cameras and cell phones with built-in cameras both became the norm.

Kodak did own many of the original patents related to the digital technology. And they made billions of dollars from those patents. Which gave the false impression that they were doing well as a company. The finite-minded leaders falsely believed that a strong balance sheet equaled a strong company. It doesn't. At least not in the context of the Infinite Game. When Kodak's patents ran out in 2007, the money dried up, and five years later Kodak filed for bankruptcy protection.

Bankruptcy is so often an act of suicide. When we look

back at the decisions that put once successful companies on a path to bankruptcy, we discover an uncomfortably high number of leaders who were obsessed with the finite game. Their Cause abandoned, instead they are left desperately clinging to business models that may have helped them become successful but could not stand the test of time. In most cases, it's not the "market conditions" or the "new technology" or any of the other stock reasons usually offered as explanations that are responsible for their company's demise. It was the leaders' inability to make the necessary Existential Flex that was the problem. If they had abandoned their Cause, they also abandoned the capacity to Flex. Call it "existential inflexibility." At some point, every single organization will need to make a Flex. Though that need might not happen during one particular leader's tenure, part of any leader's responsibility is to build their organization with the capacity to exercise Existential Flexibility should they or their successors need to do so. That means adhering to the Just Cause as a guiding light and maintaining a culture rich with Trusting Teams.

The opening sentence of the January 19, 2012, announcement of Kodak's bankruptcy in *The New York Times* summed it up perfectly: "Eastman Kodak, the 131-year-old film pioneer that has been struggling for years to adapt to an increasingly digital world, filed for bankruptcy protection early on Thursday." A statement from the chief financial officer, Antoinette McCorvey, revealed the very finite game Kodak's leaders were playing. "Since 2008," the statement went, "despite Kodak's best efforts, restructuring costs and recessionary forces have continued to negatively impact the company's liquidity position." The leaders of a once great infinite company had abandoned their moral responsibility to advance

Eastman's Just Cause in favor of a perceived responsibility to more finite ambitions. They allowed forces of the market and not a passion for the vision to dictate the company's future. And the entire company, the people who worked there, the town of Rochester, and their shareholders had to pay the price.

These days, Kodak exists as a shadow of its former self. At the time Kodak invented the digital camera, the company employed about 120,000 people. Now they employ about 6,000 people. Though the company still makes film and all the products that go with processing film, ironically its entire business primarily services one market today: professional photographers, the final nail in the coffin. Kodak had completely abandoned their founding Cause.

Without a Just Cause to guide them, Kodak's executives lacked the vision or courage to know what to do for the long-term success of their company. The most they could do was react to the world around them. George Eastman literally invented mass-market photography. The people who worked at Kodak were pioneers in almost every part of the industry. It was only their finite mindset that left this once great company to be disrupted by the visionary technology they themselves invented.

Chapter 11

THE COURAGE TO LEAD

H anging in the lobby of its corporate headquarters was a huge sign that stated their Just Cause: *Helping people on their path to better health.* And the company's executives believed it. They saw their company as having a purpose beyond just making money; they wanted to use their company to advance something bigger. They regularly had meetings with health-care companies, hospitals and physicians on how they could better work together for patients. However, near the end of many of these meetings someone would point to the elephant in the room: "But don't you sell cigarettes in your stores?"

In February 2014, CVS Caremark announced that it would stop selling any tobacco-related products in all of their over 2,800 stores. It was a decision that would cost the company $2 billion per year in lost revenue. It was a decision they chose to make even though there was no competitive pressure to do so. There was no loud public demand that they make the decision. There was no scandal. There was no online campaign to force them to make the decision.

The news was met with overwhelming support from the general public. But Wall Street and its pundits were none too pleased. "It might make money in Oz," said Jim Cramer, one of CNBC's financial commentators, "but Wall Street is not Oz. [Wall Street isn't] saying. 'You know what? I am going to buy CVS because they are good citizens.'" Cramer went on, "I'm . . . trying to figure out the earnings per share. And the earnings per share for CVS just got worse."

Other outside commentators agreed and saw the decision as a boost for CVS's competitors. One Illinois-based sales and marketing consultant pointed out that the decision translated into seven hundred packs of cigarettes a week per store that would now be sold by some other retailer, adding that "retailers know that winning the adult tobacco consumer generates incremental sales from ancillary purchases during the same visit." Looking through the lens of finite and infinite games, I can't help but see these responses to CVS's decision as exquisitely finite minded. If the game of business was a finite game and the future was easy to predict, the pundits would have been 100 percent correct. As it turns out, however, the game is infinite and the future is quite unpredictable.

In reality, that seven hundred packs of cigarettes per week per store didn't just go somewhere else. They went nowhere. The total sale of cigarettes actually decreased. An independent study commissioned by CVS to see the impact of their decision showed that overall cigarette sales dropped by 1 percent across all retailers in the states where CVS had a

15 percent market share or greater. In those states, the average smoker bought five fewer packs of cigarettes, which totaled 95 million fewer packs sold over an eight-month period. On the other hand, the number of nicotine patches sold increased by 4 percent in the period immediately after CVS stopped selling cigarettes, indicating that CVS's decision actually encouraged smokers to quit. As for the lost revenue, other purpose-driven companies who previously refused to do business with CVS also took notice. Companies like Irwin Naturals and New Chapter vitamins and supplements, whose products are available at Whole Foods and other specialty health stores, finally agreed to allow CVS to carry their products too. A move that allowed CVS to offer a greater selection of high-quality brands to their customers and open new sources of income. When a company with the stated Cause of helping people live healthier lives made a courageous decision to deliver on that purpose, not only did it help make Americans a little healthier, but it also had a positive impact on overall sales at their pharmacies.

Of course, there are many other factors that have contributed to CVS's (which soon after the decision changed their name to CVS Health) stock performance. But financial health in the Infinite Game is again, like exercise, impossible to measure in daily steps. It is a steady buildup that, in time, yields dramatic results. Jim Cramer adroitly pointed out that Wall Street isn't going to buy a company because they are good citizens. But customers and employees do. And more loyal customers and more loyal employees tend to translate into more success for the company. And the more successful a company, the more shareholders tend to benefit. Or am I missing something?

Indeed, as Cramer and other analysts predicted, CVS's

stock price did fall 1 percent the day after the announcement, from $66.11 to $65.44 per share. Only to recover the very next day. A year and a half after the announcement and eight months after the plan was implemented, the stock hit $113.65 per share, double what it had been before the announcement— and a record high for the company. And what of that "gold standard" of public company financial metrics that Jim Cramer was so worried about—the earnings per share? Prior to the announcement in December 2013, CVS had an EPS of $1.04. After the announcement it dropped to $0.95. By the next quarter it was back up to $1.06 and then rose by 70 percent to average $1.77 over the course of the next three years.

Adopting an infinite mindset in a world consumed by the finite can absolutely cost a leader their job. The pressure we all face today to maintain a finite mindset is overwhelming. For most of us, almost any kind of career opportunities we have are almost all tied to how well we perform in the finite game. Add the steady drumbeat of the analyst community, pressure from private equity or venture capital firms, the tying of executive pay packages to stock performance rather than company performance (which amazingly don't always align), our egos and the pressure many of us put on ourselves because we falsely tie our own value or self-worth to how we perform in the finite game, and any hopes we may have to do anything other than play with a finite mindset seem completely dashed. Bowing to the pressure of the finite players around us is the easy and expedient choice. This is why it takes courage to adopt an infinite mindset.

The Courage to Lead is a willingness to take risks for the good of an unknown future. And the risks are real. For it is much easier to tinker with the month, the quarter or the year, but to make decisions with an eye to the distant future is much more difficult. Such decisions may indeed cost us in the short term. It may cost us money or our jobs. It takes the Courage to Lead to operate to a standard that is higher than the law—to a standard of ethics. And when we are pressured to do things that violate that ethical code, it takes the Courage to Lead to speak up, to make those who would pressure us to do otherwise aware of the situation they are creating. And it takes courage to offer our help so they may fix it. It takes the Courage to Lead to make decisions counter to the current standards of business and it takes the Courage to Lead to ignore the pressure of outside parties who are not invested in or believers in our Just Cause.

Courage, in the Infinite Game, is not solely about the actions we take. Even leaders who operate with a finite mindset can take risks. Courage, as it relates to leading with an infinite mindset, is the willingness to completely change our perception of how the world works. It is the courage to reject Milton Friedman's stated purpose of business and embrace an alternative definition. When we have the courage to change our mindset from a finite view to a more infinite view, many of the decisions we make, like CVS's choice to stop selling cigarettes, seem bold to those with a more traditional view of the world. To those who now see the world through an infinite lens, however, such a decision is, dare I say it, obvious.

So how are we to find the courage to change our mindset?

1. We can wait for a life-altering experience that shakes us to our core and challenges the way we see the world.
2. Or we can find a Just Cause that inspires us; surround ourselves with others with whom we share

common cause, people we trust and who trust us;
identify a Rival worthy of comparison that will push
us to constantly improve; and remind ourselves that
we are more committed to the Cause than to any
particular path or strategy we happen to be following
right now.

The first method is completely legitimate and indeed is the
way so many of our great leaders came to be infinite minded.
Be it tragedy, opportunity or divine intervention—something
pushed them, sometimes quite suddenly, to see the world in
an entirely new way. This method is, however, a bit of a
gamble. . . . I would not recommend that we simply go about
our days waiting for this to happen.

The second method offers us a little more control. All
that is required is a little faith, a little discipline and the willingness to practice. For many, that conversion can feel profound. Beyond how it feels, however, such a mind shift does
indeed affect the decisions and actions we take. To those who
still see the world through a finite lens, our actions may seem
idealistic, naïve or stupid. To those who believe what we believe, our actions will seem courageous. To the infinite-
minded players out there, those courageous choices become
the only options available.

The Power of Purpose

"She told me I couldn't let our airline fail because of the dramatic impact it would have on her life as a single working
mom," he remembers.

Doug Parker was named the new CEO of America West
Airlines on September 1, 2001. Ten days later, the events of

September 11 unfolded. Though many businesses suffered, the impact on the airline industry was especially hard. U.S. passenger loads declined over the following two years, to a level that hadn't been seen since World War II. Companies like United and US Airways filed for bankruptcy protection. And for a smaller regional airline like America West, who didn't have the kind of revenue cushion the bigger airlines had, it looked like the company was going to completely collapse.

Parker was one of the first to apply for a government loan from the newly formed Air Transportation Stabilization Board (ATSB), which offered $10 billion in loans to the airline industry after September 11. But the meeting didn't go well. Flying home on an America West flight, Parker felt dejected. "It didn't look good," he recalls. "As the newly minted CEO of America West, I was going to have the shortest, least successful CEO career in history." To take a break from thinking about the day, he decided to go to the galley to talk to the flight attendants. And that's when he met Mary. An outstanding flight attendant, Mary's job meant everything to her. It was no fault of hers that she worked for an airline without the strength to survive the industry shakeout. "The only hope she had to avoid a serious personal crisis," Parker recounts, "was for the people she worked for to figure out how to keep her company afloat."

Before meeting Mary, avoiding collapse was a business matter for Parker—it was about figuring out the numbers to keep the company in business. It was only about managing the resources. After meeting Mary, it became a personal mission; it was about will too. "That commitment to a purpose bigger than ourselves drove us to accomplish things we likely would not have been able to if we were simply working on our own account," Parker explained. With newfound passion, the new CEO and his team fought for and received the government loan that had seemed impossible to get on the flight home. Looking to further strengthen the company and get a more competitive route network, Parker led America West into a merger with US Airways in 2005 and with American Airlines in 2013. "At that point the mission was accomplished," says Parker with pride. "American is the largest airline in the world. Our team was finally safe."

But something still felt off for Parker. "By early 2016, I found myself questioning my purpose for working," he said. "We had delivered on a purpose bigger than ourselves, and I was still showing up to work, but it didn't feel very fulfilling. Was I just working for money?" he remembers asking himself. "For prestige? I sure didn't like the thought of answering either of those questions in the affirmative." Parker started to ask himself if he should leave the company. Move on to do something that "could better fulfill my desire to work for a cause bigger than myself," as he put it. This is so common among highly successful people. After they finish their careers, they go on to start foundations or distribute their wealth to charity, working to fulfill a desire to give back, do that "something" more philanthropic. But purpose is not something we only find after a successful career.

Parker's drive to serve Mary and her colleagues, though incredibly inspiring, was framed as a moon shot. It had an end point. And once completed, Parker was left searching again. He had tasted what it felt like to be driven by something bigger than himself. It ignited a passion in him to drive the company to succeed like never before—not for his own glory, but for others. And he wanted that feeling again.

Parker heard a talk given by Bob Chapman, the CEO of

the manufacturing company Barry-Wehmiller. Chapman (about whom I wrote extensively in *Leaders Eat Last*) is an outspoken voice for the idea that the best leaders and the best companies prioritize people before numbers. That his company consistently thrives beyond expectations with a people-before-profit philosophy earns him invitations to speak to the converted and skeptics alike. It was at one of those talks that Parker was struck with a clear realization—he recognized the moon shot but he hadn't yet recognized the context for that moon shot. Working to give people job security and higher pay may be an essential milestone on his journey, but it wasn't the Just Cause that could inspire him for the rest of his life. "We needed to create an environment that cared for them! Where they were recognized and appreciated for their great work; where their leaders cared about them; and where they went home at the end of the day feeling fulfilled. That was the new mission bigger than myself I'd been looking for," Parker says excitedly about his new infinite pursuit.

So what happens when the CEO of the largest airline in the world has the courage to change how he leads—to move from a finite to an infinite mindset?

Like so many companies that prioritize numbers before people, American Airlines had a history of trust issues with employees. Long before Doug Parker showed up, the previous leadership team had negotiated significant concessions from the unions in the name of "helping the company manage bankruptcy protection," while at the same time, guarantees were made to the top seven executives that they would receive bonuses worth double their salaries simply to stick around for a few more years. As if that wasn't bad enough, $41 million was put aside to protect the pensions of the top

45 executives. And no such provisions were made for rank-and-file employees.

The scandal ultimately resulted in the resignation of then CEO Donald Carty. In his departing statement, he expressed hope that his successors would try to build a "new culture of collaboration, cooperation and trust." Something that, despite public assurances, his successors, Gerard Arpey and Tom Horton, were unable to do. And the trust violations and possible ethical fading persisted. Unless a new leadership team was willing to make some hard choices and some sacrifices to demonstrate that they were indeed worthy of trust, nothing was going to change. Parker understood that grand pronouncements of how things were going to be different would do little to move the needle. He knew he and his leadership team needed to find the courage to demonstrate that things were, in fact, going to be different. And that's exactly what they did.

Their first significant act happened in 2015, when they negotiated new contracts for their pilots and flight attendants that would make them some of the best paid in the industry. A year later, however, Delta and United signed new pilot and flight attendant contracts of their own, leapfrogging American by 5 percent for flight attendants and 8 percent for pilots. With the culture of cynicism still alive and well, many believed, falsely, that leadership knew this would happen and worked to hurry up and lock them into the lower contracts for the next five years.

"Saying you trust people is just words," says Parker. "To validate the trust, we have to act in a way that lives up to the words." A lot of other executive teams would simply shrug and promise to deal with it at the next contract negotiation.

"Isn't that the purpose of a contract?" they may say. However, trust is not built by pressure or force, trust is built by acting in a way consistent with one's values, especially when it's least expected. Trust is built when we do the right thing, especially when we aren't forced to. And seeing their employees left behind industry averages for three or four more years just didn't "feel right for the new American and it doesn't feel consistent with our commitment," according to a joint statement issued by Parker and company president Robert Isom.

The senior executives decided to give all their flight attendants and all their pilots a midcontract raise of 5 percent and 8 percent, respectively, and asked for nothing in return. The decision would cost the company over $900 million over the next three years. It was a decision they knew Wall Street would hate. And they were right.

On April 27, 2017, when American made the announcement, Wall Street's reactions were predictably disapproving. One analyst, Kevin Crissey, who specializes in the airline industry for Citi, wrote to his clients, "This is frustrating. Labor is being paid first again. Shareholders get leftovers." A letter from a group of J.P. Morgan analysts echoed the sentiment. "We are troubled by AAL's wealth transfer of nearly $1 billion to its labor groups," said the opening line of the letter. "We're sensitive to American's desire to 'build a foundation of trust' with its labor stakeholders," the letter later explained, "but we think this latest agreement goes too far. . . . The solution to a rising wage bar is not to chase it, in our view. Sometimes, the timing of one's commitments is simply fortuitous." By "fortuitous," I believe they were saying "possibly unfair, but in our favor." Fortunately, the leaders at American Airlines had the courage to make a decision to strengthen their company without considering Mr. Crissey's and the J.P. Morgan team's annual bonus structures.

Sadly, it is finite mindsets of those like Mr. Crissey and the analysts at J.P. Morgan who help sway the market. American predicted they would lose as much as 5 percent of their stock value. The day after the announcement, the stock price actually lost 9 percent of its value. The good news is, short-term thinking often has short-term impact. In less than two weeks the stock regained its full original value and, by year end, was over 20 percent higher. Even so, many on Wall Street argue that American would be more profitable if it hadn't given its employees raises. Once again, demonstrating their bias for resources over will. Finite thinkers do not appreciate that an investment in people will ultimately benefit the company, the customer and their investments (and they probably also fail to recognize that it was their guidance that pushed the stock price down).

A CEO of a major public company pointed out to me that Wall Street analysts tend to write for the short-term community. So they tend to write the things that promote their interests—finite objectives. Responding to a question about all that short-term analyst chatter, Parker admitted that it is hard to completely ignore. We "have to work on it, we can quickly get sucked into it," he said. The good news is, Parker, his team and the board of directors are working hard to be less reactive to the noise and to stay focused on the long term. "We need to care for our team so that they can care for our customers," said Parker. "That's how we will create value for our shareholders."

American Airlines is still in the early days of their new journey. But because they are now preaching more of a long-term story than in the past, they are, unsurprisingly, attracting the attention of more long-term-minded investors. The kinds of investors who care less about the short-term fluctuations. One of those investors is Ted Weschler. Weschler is one of four investment managers who run Warren Buffett's Berkshire Hathaway, a company well known for their long-term positions; they rarely sell off their investments. (As it turns out, long-term shareholders, like Berkshire Hathaway, have their own analysts and tend not to be swayed by the twenty-four-hour financial news cycle.)

Buffett—the Oracle of Omaha, one of the most successful investors in history, one of the richest men in the world, revered in financial circles worldwide—once wrote that airlines were one of the worst investments someone could make. As he explained in a 2007 Berkshire Hathaway shareholder letter, "The worst sort of business is one that grows rapidly, requires significant capital to engender the growth, and then earns little or no money. Think airlines. Here a durable competitive advantage has proven elusive ever since the days of the Wright Brothers. Indeed, if a farsighted capitalist had been present at Kitty Hawk, he would have done his successors a huge favor by shooting Orville down." It's worth noting that, at the time of the publication of this book, Berkshire Hathaway is the single largest shareholder of American Airlines. And when Doug Parker informed them of his intention to give the midcontract raise to his flight attendants and pilots, Weschler gave Parker his blessing. The joke is, all those finite thinkers who complain about Parker's leadership perspective will probably still invest in American if they think they can make a buck.

It Takes No Courage to Keep a Finite Mindset

CVS decided to use their Just Cause to guide their business and they were the first to take the risk to remove cigarettes from their stores. This should make it easier for others to follow their lead. However, as of the writing of this book, its two biggest competitors, Walgreens and Rite Aid, continue to stock cigarettes on their shelves. I wanted to give them the benefit of the doubt. Even though they are both pharmacies, perhaps Walgreens and Rite Aid chose to stay the course because they have a different cause than CVS. Perhaps their decisions are consistent with their stated purposes. So I checked, to be sure.

On the "About us" section of the Walgreens Boots Alliance (the company that owns Walgreens pharmacies) website, it states that its purpose is to "help people across the world lead healthier and happier lives." After which it states, "Walgreens Boots Alliance takes seriously its aim of inspiring a healthier and happier world, as reflected in our core values." Of which the first is, "Trust: respect, integrity and candor guide our actions to do the right thing." When asked if they plan to follow CVS's lead, Walgreens released a statement that included their "active decision to reduce space and visibility of tobacco products in certain of our stores as we focus on helping customers who want to stop smoking." Bold, Walgreens, bold.

Executive chairman of Walgreens Boots Alliance James Skinner responded to the same question by stating, "We've reviewed this on a regular basis and it's always up for a review decision down the road." Isn't that the opposite of courage or conviction? What exactly is Mr. Skinner afraid will happen if he makes a decision consistent with the company's actual stated purpose?

costs of smoking

According to the Centers for Disease Control (CDC), smoking is the leading preventable cause of death in the United States. The number of people who die from smoking-related illness each year is greater than all the people who die from HIV, illegal drug use, alcohol use, car accidents and firearm-related incidents combined! Cigarettes kill 480,000 people *every single year*. That's 80,000 more than the total number of American servicemen who died in *all* of World War II! The economic costs are also exorbitant. All those smoking-related illnesses cost American taxpayers more than $300 billion *each year*. The entire cost of NASA's Space Shuttle program, which includes building six space shuttles (five of which flew to space), cost taxpayers $196 billion over the course of more than thirty years (an average of $6.5 billion per year). The *annual* health-care total related to smoking costs the country nearly fifty times more than traveling to space!

If an oil company is held responsible for the costs associated with an oil spill or even a leaky pipeline, if car companies are held responsible when defects in a car's design causes injury, then shouldn't tobacco companies and the stores that sell their products be held responsible for that $300 billion annual cost? Remember those errors in causal perception in the ethical fading section. *Of course* a pharmacy devoted to helping people be healthy that sells a highly addictive and cancer-causing product like cigarettes bears *some* responsibility for the ill health they cause their customers, yes?

The single best way to prevent all the deaths and reclaim all the money we lose to smoking-related illness is to help smokers stop smoking. Something most smokers *want* to do. Nearly 70 percent of all smokers, many of whom shop at pharmacies, report a desire to quit. But it's not easy and, obviously, many struggle to do so. Which is why offering them an antismoking program next to the cigarettes isn't much help. That's a little like selling doughnuts next to diet books. The choice facing the consumer is between one item that satisfies a craving and is bought on impulse and another that requires discipline and hard work. Anyone who *actually* wanted to help would try to make the harder choice a little easier by fully removing the thing that drives the impulse . . . even if there is a cost to doing so. That's what the Courage to Lead is!

If leaders of organizations go so far as to state a Just Cause, or purpose, for their organization, then it's kind of necessary that they must actually believe in that Cause. The whole point of having a statement of Cause or purpose is that they *actually* believe it. That they *really* believe the purpose of business is bigger than making money. A Cause can only advance if they do the things that help advance it. If they don't, what's the point of having a Cause written on the wall or on the website?

More and more people say they want to work for a purpose-driven organization, especially Millennials and Gen Zers. But without committed, infinite-minded leaders willing to challenge accepted norms of how the working world works, statements of Cause are just feel-good marketing—stuff a company may say to curry favor with people inside or outside the organization, but may not actually believe in or do themselves. Perhaps the pressure to make their numbers is acting on business leaders like the seminary students at Princeton. If the leaders of companies have no real interest adopting an infinite mindset or at least being open to the idea that maybe they don't have everything figured out, they can at least have the courage to say what their true intentions are and delete from their websites and marketing what appear to be hollow statements of purpose or cause. Being honest about their short-term intentions would be, as Walgreens explains in its values, operating with integrity in order to build trust. But alas . . . that too takes courage.

After the CVS announcement, Rite Aid, the third of the big three pharmacy chains in America, responded to the same questions about whether it would follow suit. After all, doing so would also be consistent with its stated purpose. The first sentence of the "our story" section of the pharmacy website reads, "At Rite Aid, we have a personal interest in your health and wellness. That's why we deliver the products and services that you, our valued customer, need to lead a healthier, happier life." Yet, when asked whether they planned to follow CVS's lead and stop selling cigarettes in their stores, the company released a statement that Milton Friedman himself could have written: "Rite Aid offers a wide range of products, including tobacco products, which are available for purchase in accordance with federal, state and local laws."

Think about that for a moment. When a company responds to an ethical question (or defends an unethical decision) by explaining that they can legally do what they are doing, that's like someone who has been caught cheating by their long-term boyfriend or girlfriend replying, "What?! We're not married. I broke no laws. I'm legally allowed to sleep with someone else if I want." Their actions may indeed be legal, but it is hardly the kind of response that engenders or rebuilds trust.

When companies and the people who lead them act with courage and integrity, when they demonstrate that they are honest and of strong character, they are often rewarded with good will and trust from customers and employees. The day after CVS made the announcement that it would be pulling cigarettes from all its stores, the phone rang on Maryalyce Saenz's desk. It was her mother. Almost in tears, she told Maryalyce how proud she was that her daughter worked for a company like CVS. For years, Maryalyce's father's smoking habit had been a source of family conflict. "That was a really gutsy move," explained Maryalyce. "I was really proud to come to work that day. And, I think out of everything," she continued, "that was the day where I sat back and I thought, 'I am absolutely in the right place.'" It's safe to say that neither employees nor customers get the same warm and fuzzy feelings when a company obeys the law.

The courage to see the Infinite Game—to see the purpose of business as something more heroic than simply making money, even if it's unpopular with the finite players around us—is hard. True Courage to Lead holds the company and its leadership to a much higher standard than simply acting within the bounds of the law. Only when organizations operate on a higher level than federal, state and local laws can we say they have integrity. Which, incidentally, is the *actual* definition of integrity—firm adherence to a code of especially moral or artistic values: incorruptibility. Indeed, the pursuit of a Just Cause is a path of integrity. It means that words and actions must align. It also means that there will be times when leadership must choose to ignore all the voices calling for the company to serve the interests of those who don't necessarily believe in the Cause at all.

Integrity does not just mean "doing the right thing." Integrity means acting before the public outcry or scandal. When leaders know about something that is unethical and only act after the outcry, that's not integrity. That's damage control. "They wait for public opinion to tell them what to do," said Rosabeth Moss Kanter, a professor at Harvard Business School, when talking about how CEOs make decisions today. "CEO courage is in short supply."

Splits and Crossroads

Human beings are messy and imperfect. There is no such thing as a perfectly infinite-minded leader and there is certainly no such thing as a perfectly infinite-minded organization. In reality, even the most infinitely focused companies can stray onto a finite path. And when that happens, it takes the Courage to Lead to recognize that the organization has strayed from its Cause and it takes courage of leadership to get back on course.

This is sadly common once an organization has achieved great success. Whereas the infinite-minded player sees that they are still at the tip of the iceberg no matter how much traditional success they enjoy, the finite player will often transition into playing defense to guard their pole position. It takes Courageous Leadership to stay in the Infinite Game after you arrive at the top. To recognize that, regardless of how much success has been achieved, the Cause is infinite. Unfortunately, the temptation to convert to finite is so, so tempting.

There was a period, for example, when the Disney corporation strayed from its infinite Cause to chase more finite pursuits like global domination, enhanced shareholder value and the enrichment of those who chose to enable it. In 1993, Disney bought Miramax Films, which went on to produce such family-friendly movies as Quentin Tarantino's crime flick *Pulp Fiction*; Danny Boyle's black comedy about Scottish heroin addicts, *Trainspotting*; and a rereleased edit of Francis

Ford Coppola's surreal ride into the Vietnam War, *Apocalypse Now Redux*. Under Disney's record label Hollywood Records, we were able to enjoy such family-friendly acts as the hard-core punk band Suicide Machines and heavy metal band World War III.

Whenever a new CEO takes over, that new leader will stand at a crossroads. How will they lead? When Mike Duke and Steve Ballmer took the helms at Walmart and Microsoft, respectively, both made the choice to lead their companies down a finite path. Had the companies stayed on these paths, they may have been forced to drop out of the game altogether. The CEOs who replaced them, Doug McMillon at Walmart and Satya Nadella at Microsoft, also made a choice—to do what they needed to do to put their respective companies back on the infinite path. And though they still face many challenges, both seem genuinely committed to leading a Cause, not just running a company.

Major events, like an IPO or change in leadership, can force an organization to choose one path over the other too. However, there need not be a specific event to cause an organization to veer from the infinite path to a finite one. Such veerings or splits off the infinite path are actually quite normal. People stray from their own paths all the time. We often stray from a healthy routine or fall off other healthy bandwagons. As companies are run by people, it would be expected that these things will happen. What causes an organization to stray off course is often quite consistent. It occurs when leaders become more interested in their own finite pursuits than the Infinite Game and drag the organization along with them.

Organizations will also find themselves at a crossroads when their leaders start to believe their own myths—that the success the company enjoyed under their leadership was a result of their genius rather than the genius of their people, who were inspired by the Cause they were leading. These leaders too often fixate on advancing their own fame, fortunes, glory and legacies at the expense of the company and its Cause. Management becomes disconnected from the people and trust breaks down. And when performance necessarily starts to suffer as a result, these same leaders are quicker to blame others than to look at what set the company on the new path in the first place. In order to "fix" the problem, their faith in the people is replaced with faith in the process. The company becomes more rigid and decision-making powers are often taken away from the front lines. It can't be a good thing when the captain of the ship, who is supposed to be on deck navigating toward the horizon, is now in the ship tinkering with the engine trying to make it go faster.

Facebook was an infinite player that now seems to be moving down a more finite path. Founded in 2004, Facebook came to life with a well-articulated Cause to "give people the power to build community and bring the world closer together." Today, however, it finds itself embroiled in scandals that do anything but "bring the world closer together." Facebook has been accused of violating their users' privacy, tracking our habits online (even when we're not on Facebook), failing to adequately police fake accounts or fake news disseminated across their service, then using all the data they collect either to sell or to maximize the dollars they can earn from selling advertising. I doubt this is what Mark Zuckerberg meant by "giving people power." Has Facebook veered from their once inspiring infinite path because of the overwhelming pressure their leaders feel to answer to Wall Street's finite expectations? Is it because they are doubling down on a business model driven by selling advertising instead of making an Existential Flex to reshape the entire company? Is it because their leaders have lost connection with their Just Cause and who they need to be primarily serving in order to keep the game in play? Is it hubris? Today, when Facebook does right by the people, it is too often a result of public pressure or scandal and rarely a proactive decision made to protect those they serve and advance their Cause. Facebook reacted to the scandal that erupted around Cambridge Analytica, for example, only after there was a scandal, even though they were aware of Cambridge Analytica's unethical practices before we found out about it. Regardless of what combination of things led Facebook down this path, there is no getting around the fact that they are acting with a more finite mindset than in the past. Being big and rich does not mean the company can't fail. Though money certainly helps delay the inevitable in this never-ending game. It also provides the runway for leaders to get things back on track. The only question is whether they will or not. With a little Courage to Lead they can renew the trust of the people who helped champion their success before it's too late.

As companies like Microsoft, Walmart and Disney show, companies can afford to veer off course for a while. They will still face the challenge of finding their way back to the infinite path that they once all started down. Though some can bear the cost of splitting for longer, money eventually runs out. Not every organization can afford to veer off the infinite path for as long. Regardless of the size of the company, the elements of infinite-minded leadership that I've tried to make a case for in this book are the best way to help stay on that infinite journey. Playing the Infinite Game is not a checklist, it's a mindset.

How to Find the Courage to Lead

In my life, the only common factor in all my failed relationships is me. The common factor in all the struggles and setbacks that finite leaders face is their own finite thinking. To admit that takes courage. To work to open one's mind to a new worldview takes even more courage. Especially when we know many of our choices will go badly. To actually take steps to apply an infinite mindset to an organization's culture can seem to many like it would take insurmountable courage. And the truth is, it does. For it can be embarrassing, even humiliating, to admit that we are part of the problem. It can also be empowering and inspiring to decide to be a part of the solution.

Few if any of us have the courage to change from a finite mindset to a more infinite one alone. We must find others who share our sense of responsibility, who share our beliefs that it is time to change and who share our desire to work together to do it. In every case I wrote about to demonstrate the Courage to Lead, the hard decisions were not made by great women and great men. They are done by great partnerships. Great teams. Great people who stood together with deep trust and common cause. Like a world-famous trapeze artist would never attempt a brand-new death-defying act for the first time without a net, neither can we find the courage to lead without the help of others. Those who believe what we believe are our net.

Courageous Leaders are strong because they know they don't have all the answers and they don't have total control. They do, however, have each other and a Just Cause to guide them. It is the weak leader who takes the expedient route. The ones who think they have all the answers or try to control all the variables. It requires less strength to announce layoffs at the end of the year to quickly squeeze the numbers to meet an arbitrary projection than it does to explore other, maybe untested, options. When leaders exercise the Courage to Lead, the people who work inside their organization will start to reflect that same courage. Like children who mirror their parents, so too do employees mirror their leaders. Leaders who prioritizes themselves over the group breed cultures of employees who prioritize their own advancement over the health of the company. The Courage to Lead begets the Courage to Lead.

AFTERWORD

Our lives are finite, but life is infinite. We are the finite players in the infinite game of life. We come and go, we're born and we die, and life still continues with us or without us. There are other players, some of them are our rivals, we enjoy wins and we suffer losses, but we can always keep playing tomorrow (until we run out of the ability to stay in the game). And no matter how much money we make, no matter how much power we accumulate, no matter how many promotions we're given, none of us will ever be declared the winner of life.

In any other game, we get two choices. Though we do not get to choose the rules of the game, we do get to choose if we want to play and we get to choose how we want to play. The game of life is a little different. In this game, we only get one choice. Once we are born, we are players. The only choice we get is if we want to play with a finite mindset or an infinite mindset.

If we choose to live our lives with a finite mindset, it means we make our primary purpose to get richer or promoted faster than others. To live our lives with an infinite mindset means that we are driven to advance a Cause bigger than ourselves. We see those who share our vision as partners in the Cause and we work to build trusting relationships with them so that we may advance the common good together. We are grateful for the success we enjoy. And as we advance we work to help those around us rise. To live our lives with an infinite mindset is to live a life of service.

Remember, in life, we are players in multiple infinite games. Our careers are just one. No one of us will ever be declared the winner of parenting, friendship, learning or creativity either. However, we can choose the mindset with which we approach all these things. To take a finite approach to parenting means to do everything we can to ensure our kids not just get the best of everything but are the best at everything. A seemingly fair standard, for these things "will help our kid excel in life." Except when a finite mindset is the primary Strategy, it can give way to ethical fading or push us to become more obsessed with our child's standing in the hierarchy over if they are actually learning or growing as a person. An extreme example is shared by clinical psychologies and *New York Times* bestselling author Dr. Wendy Mogel. She tells the story of a father who raised his hand during a conference at which she was speaking to tell her that "he had a fight with the pediatrician about his son's apgar score . . . and I won." The apgar score is a test performed within the first minute to five minutes of a child's birth to determine their strength. Basically, as Dr. Mogel explains, "if they are blue and floppy, you get a one, if they are pink and plump they get a five." Think about that for a second. This parent seemed more concerned with "winning" and getting his newborn child a higher score rather than concerning himself with his child's health. Flash forward 18 years and think about the lengths that parent might go to ensure his child gets the best scores to get into the best school all the time ignoring if their child is actually learning or is healthy in every other way.

To parent with an infinite mindset, in contrast, means helping our kids discover their talents, pointing them to find their own passions and encouraging they take that path. It means teaching our children the value of service, teaching them how to make friends and play well with others. It means teaching our kids that their education will continue for long after they graduate school. It will last their entire lives . . . and there may not be any curriculum or grades to guide them. It means teaching our kids how to live a life with an infinite mindset themselves. There is no single, greater contribution in the Infinite Game than to raise children who will continue to grow and serve others long after we are gone.

To live a life with an infinite mindset means thinking about second and third order effects of our decisions. It means thinking about who we vote for with a different lens. It means taking responsibility for later impact of the decisions we make today.

And like all infinite games, in the game of life, the goal is not to win, it is to perpetuate the game. To live a life of service.

None of us wants on our tombstones the last balance in

our bank accounts. We want to be remembered for what we did for others. Devoted Mother. Loving Father. Loyal Friend. To serve is good for the Game.

We only get one choice in the Infinite Game of life. What will you choose?

■ ■ ■

If this book inspired you, please pass it on to someone you want to inspire.

ACKNOWLEDGMENTS

Ideas evolve. They are not like a light that is suddenly turned on with a switch. Nor are they random. We have ideas about questions that have been raised or problems that we are grappling with. And if there is an ah-ha moment, it comes only after we've been reading things, watching things, listening to things and having conversations with others—all things that contribute to, inspire and point us in a direction that our ideas may form. This was certainly the case for *The Infinite Game*.

The seed for this book was planted years ago when my friend, Brian Collins, gave me a copy of James Carse's book, *Finite and Infinite Games* (thank you Dr. Carse for writing that magical little book). I became enamored by the idea and it started to influence the way I saw the world. I subsequently gave dozens of copies away to others who I thought would appreciate the alternative perspective. One of those people was Andy Hohen at RAND Corp. Andy and I had many long conversations about how the idea of the Infinite Game was a new lens through which to view global politics and military strategy. David Shedd, a big thinker and long-time public servant, challenged me with hard questions which further helped shape my thinking. I was lucky to be invited by Brig. Gen. Blane Holt, USAF (ret.) and Mike Ryan, SES, to attend a EuCom gathering in Germany where I had the chance to share ideas on how we can use an infinite mindset to better understand America's role in a post-Cold War world. Then, at an entrepreneurial conference in New York City, Seth Godin gave a talk that inspired me to abandon my script to try something new. That was the first time I applied the infinite mindset to business. It became clear that we needed more than a new lens through which to view the world—we needed to understand what it meant to lead in a world in which most, if not all, of us were playing in an infinite game of some sort or another.

As the idea started to grow, I needed to test it. There were some early adopters who took the risk to let me share my still-forming idea in front of live audiences. Bob Patton from EY let me talk about it at his company's Strategic Growth Forum in Palm Springs, California. TED gave me an audience in New York. Google let me hash it out with their folks too. And William Morris Endeavor encouraged me challenge their leadership with what it means to lead with an infinite mindset. And slowly but surely, the ideas of how to lead in the Infinite Game took greater form. To all the people who jostled with me and gave me a chance to test the ideas to real audiences, of which there are so many, thank you.

When I finally brought the idea to my publisher, Adrian

Zackheim, as he has done with me in the past, he smiled and said, "I'll publish that." A deep heartfelt thank you to Adrian for taking yet another bet on one of my nutty ideas on how I think the world could work. And then the real work began— the work of writing the book.

Writing a book is a combination of research and writing, more conversations and debates, then refining and rewriting. It is filled with all the emotions . . . ALL of them. And the one person who stood with me as I went through all those emotions is Jenn Hallam. My compadre from the get go, you pushed me to make my ideas stronger, you helped me make the writing clearer. I could not have written this book without you. Jenn . . . thank you, more than I have words, thank you.

While I was deep down the rabbit hole of writing, my team picked up the slack. To Sara Toborowsky, Kim Harrison, Lori Jackson, Melissa Williams, Molly Strong, Monique Helstrom and Laila Soussi and the rest of my team—thank you for being so patient with me and taking care of me and everything else that needed taking care of for all those months.

A special thank you to Tom Staggs for the hours and hours you gave me to help make the ideas and this book stronger. I so value your counsel and friendship. Thank you Lt. Gen George Flynn, USMC (ret.), you were by my side through the whole journey—tinkering with me when it was an outline to reading the final manuscript—thank you. Thank you to Tom Gardner and the folks at Motley Fool for sharing your vast knowledge. Thank you to Adam Grant, my Worthy Rival and friend. You are so good at what you do— you inspire me to be better. To Bob Chapman, my partner in Cause. Our torch is burning brighter and brighter each day.

To the whole gang from STRIVE Morocco, thank you. It was with you in the desert that I was inspired to talk about what it means to live an infinite life for the first time (it may have had something to do with how I felt after riding up that hill earlier that day).

To the people who shared their thinking and their stories with me to bring this book to life: Angela Ahrendts, Christine Betts, Chief Jack Cauley, Officer Jake Coyle and all the wonderful people I met at CRPD, Sasha Cohen, John Couch, CAPT Rich Diviny, USN (ret.), Carl Elsener, Jeff Immelt, Curtis Martin, Steve Mitchell, Alan Mulally, Doug Parker, Joe Rohde, Maj. William Swenson, USA and Lauryn Sargent and Scott Thompson, thank you. A special thank you to Kip Tindell, for more than your stories, but for your belief in me and your encouragement.

To those who opened their minds, then challenged and pushed me—Sara Blakely, Linda Boff, Gen. Kevin Chilton, USAF (ret.); Col. Mike Drowley, USAF; Elise Eberwine; Al Guido; Brian Grazer; David Kotkin; Capt. Maureen Krebs, USMC; Jamil Mahoud; Cmdr C.K. Morgan, USN from HSM-51 (you don't know this, but your thank-you letter reframed my whole outline); Essie North; Maj. Gen. David Robinson, USAF (ret.); Gen. Lori Robinson, USAF (ret.); Daisy Robinton; Craig Russell; Jen Waldman; Kevin Warren; Mike Wirth—from the bottom of my heart, thank you.

To the leaders, at every rank, from the United States Air Force, Army, Coast Guard, Navy and Marine Corps who tested my mettle, thank you.

And most of all, the biggest thank you goes to you, the reader. To those who have joined me in this Just Cause. It is my honor to serve you as we work together to build a world in which the vast majority of people wake up inspired, feel safe at work and return home fulfilled at the end of the day. Inspire on!

NOTES

INTRODUCTION

2 **North Vietnam lost:** *The Fog of War: Eleven Lessons from the Life of Robert S. McNamara*, directed by Errol Morris (Los Angeles: Sony Pictures, 2003), www.errolmorris.com/film/fow_transcript .html.

CHAPTER 1: FINITE AND INFINITE GAMES

6 **For years, British Airways:** Janet Guyon, "British Airways Takes a Flier," September 27, 1999, archive.fortune.com/magazines /fortune/fortune_archive/1999/09/27/266152/index.htm.

8 **Though he knew it wouldn't be easy:** Daniel Eran Dilger, "Microsoft Abandons Zune Media Players in Defeat by Apple's iPod," March 14, 2011, Apple Insider, appleinsider.com/articles /11/03/14/microsoft_abandons_zune_media_players_in_ipod _defeat.

9 **Their drive is not to beat the quarter:** Jonathan Ringen, "How Lego Became the Apple of Toys," *Fast Company*, January 8, 2015, www.fastcompany.com/3040223/when-it-clicks-it-clicks.

10 **In 1912, Kodak was the first:** Rick Wartzman, *The End of Loyalty: The Rise and Fall of Good Jobs in America* (New York: PublicAffairs, 2017), 20–21.

13 **In good times, Victorinox:** Epoch Times staff, "Staying True to Values: Interview with Carl Elsener Jr., Victorinox CEO," *Epoch Times*, August 8, 2016, www.theepochtimes.com/staying-true-to -values-interview-with-carl-elsener-jr-victorinox-ceo_2132648.html.

14 **"You must never have read":** *The Fog of War: Eleven Lessons from the Life of Robert S. McNamara*, directed by Errol Morris (Los Angeles: Sony Pictures, 2003), www.errolmorris.com/film/fow _transcript.html.

16 **Debuting with a 9 percent market share:** Tim Beyers, "Too Zune for Hype," *Motley Fool*, November 20, 2006, www.fool .com/investing/value/2006/11/30/too-zune-for-hype.aspx; and Dan Frommer, "Apple iPod Still Obliterating Microsoft Zune," *Business Insider*, July 12, 2010, www.businessinsider.com/through-may -apples-ipod-had-76-of-the-us-mp3-player-market-while -microsofts-zune-had-1-according-to-npd-gro-2010-7.

17 **Spanx, Sriracha, and GoPro:** Meg Prater, "9 Brands that Survive Without a Traditional Marketing Budget," *HubSpot*, July 17, 2017, blog.hubspot.com/marketing/brands-without-traditional -marketing-budget.

17 **Yet, within four years:** Rachel Rosmarin, "Apple's Profit Soars on iPod Sales," *Forbes*, July 19, 2006, www.forbes.com/2006/07/19 /apple-ipod-earnings_cx_rr_0719apple.html#4e7d9a357f6c.

18 **Questioned about the iPhone:** Jay Yarrow, "Here's What Steve Ballmer Thought about the iPhone Five Years Ago," *Business Insider*, June 29, 2012, http://www.businessinsider.com/heres-what -steve-ballmer-thought-about-the-iphone-five-years-ago-2012-6.

19 **after just five years on the market:** Kurt Eichenwald, "Microsoft's Lost Decade," *Vanity Fair*, August 2012, www.vanityfair.com /news/business/2012/08/microsoft-lost-mojo-steve-ballmer.

19 **"In the last five years":** Mary Jo Foley, "For Steve Ballmer, a Lasting Touch on Microsoft," *Fortune*, December 10, 2013, fortune .com/2013/12/10/for-steve-ballmer-a-lasting-touch-on-microsoft.

20 **Microsoft became obsessed:** Matt Weinberger, "How Microsoft CEO Satya Nadella Did What Steve Ballmer and Bill Gates Couldn't," *Business Insider*, January 30, 2016, www.businessinsider .com/satya-nadella-achieved-one-microsoft-vision-2016-1.

21 **"lean competition machine":** Kurt Eichenwald, "Microsoft's Lost Decade," *Vanity Fair*, July 24, 2012, www.vanityfair.com /news/business/2012/08/microsoft-lost-mojo-steve-ballmer.

22 **According to a study by McKinsey:** Stéphane Garelli, "Why You Will Probably Live Longer Than Most Big Companies," IMD, December 2016, www.imd.org/research-knowledge/articles/why

-you-will-probably-live-longer-than-most-big-companies; www
.mckinsey.com/business-functions/strategy-and-corporate-finance
/our-insights/reflections-on-corporate-longevity.

22 **And according to Professor Richard Foster:** Kim Gittleson,
"Can a Company Live Forever?," BBC News, January 19, 2012,
www.bbc.com/news/business-16611040.

22 **After the 1929 stock market crash:** Simon Sinek, *Leaders Eat
Last* (New York: Portfolio/Penguin, 2017) (Glass-Steagall and
Stock Market Crashes).

CHAPTER 2: JUST CAUSE

28 **"A child died":** Volker Wagener, "Leningrad: The City That
Refused to Starve in WWII," DW.com, August 9, 2016, p.dw
.com/p/1JxPh.

29 **"I would like the Department":** Carolyn Fry, *Seeds: A Natural
History* (Chicago: University of Chicago Press, 2016), 30–31.

31 **"We shall go into the pyre":** Jules Janick, "Nikolai Ivanovich
Vavilov: Plant Geographer, Geneticist, Martyr of Science,"
HortScience 50, no. 6 (June 1, 2015): 772–76.

31 **"It was hard to walk":** Gary Paul Nabhan, *Where Our Food
Comes From: Retracing Nikolay Vavilov's Quest to End Famine*
(Washington, D.C.: Shearwater, 2009), 10.

36 **Marie Haga, the executive director:** Michael Major, "The
Vavilov Collection Connection," Crop Trust, March 19, 2018,
www.croptrust.org/blog/vavilov-collection-connection.

36 **Vizio, the California-based maker:** "Irvine California
Jobs," Vizio, careers.vizio.com/go/Irvine-California-Jobs
/4346100.

CHAPTER 3: CAUSE. NO CAUSE.

52 **"We choose to go to the moon":** President John F. Kennedy,
Moon speech, Rice University, Houston, Texas, September 12,
1962, NASA, er.jsc.nasa.gov/seh/ricetalk.htm.

52 **More than an ideal future state:** Jim Collins, *Good to Great:
Why Some Companies Make the Leap . . . and Others Don't*
(New York: HarperCollins, 2001), and Jim Collins and Jerry I.
Porras, *Built to Last: Successful Habits of Visionary Companies*
(New York: HarperCollins, 2004).

53 **Jack Welch, then CEO:** Quote comes from the author's interview
with Jeff Immelt. We need to go back to see if the topic of the
town halls came from him too. Though I fear it didn't.

54 **"We will be the global leader":** Garmin, "About Us," "Our
Vision," www.garmin.com/en-US/company/about.

57 **Imagine you walk out:** going on vacation metaphor is derived
from the work of sales coach Jack Daly.

58 **Especially in the start-up world:** Eric Paley, "Venture Capital Is a Hell of a Drug," *Tech Crunch*, September 16, 2016, techcrunch .com/2016/09/16/venture-capital-is-a-hell-of-a-drug.

58 **For companies in those markets:** Robert J. Samuelson, "Capitalism's Tough Love: The Real Lessons from the Fall of Sears and GE," *The Washington Post*, January 13, 2019, www .washingtonpost.com/opinions/capitalisms-tough-love-the-real -lessons-from-the-fall-of-sears-and-ge/2019/01/13/fef2d576-15df -11e9-803c-4ef28312c8b9_story.html.

58 **"70%–90% of acquisitions":** Roger L. Martin, "M&A: The One Thing You Need to Get Right," *Harvard Business Review*, June 2016, hbr.org/2016/06/ma-the-one-thing-you-need-to-get-right.

CHAPTER 4: KEEPER OF THE CAUSE

61 **"If we work together":** Barbara Farfan, "Overview of Walmart's History and Mission Statement," *The Balance Small Business*, July 25, 2018, www.thebalancesmb.com/history-of-walmart-and -mission-statement-4139760.

62 **"[Walmart] is very well positioned":** "Mike Duke Elected New Chief Executive Officer of Wal-Mart Stores, Inc.," Walmart, November 21, 2008, corporate.walmart.com/_news_/news-archive /investors/mike-duke-elected-new-chief-executive-officer-of-wal -mart-stores-inc-1229111.

63 **There was also a congressional investigation:** Josh Eidelson, "The Great Walmart Walkout," *The Nation*, December 19, 2012, www.thenation.com/article/great-walmart-walkout.

63 **What happened at Walmart:** Simon Sinek, "Why Too Many Successions Don't Succeed," *Huffington Post*, December 27, 2008, www.huffingtonpost.com/simon-sinek/why-too-many -successions_b_146700.html.

65 **"I will go up and out":** From a conversation with General Lori Robinson.

66 **"I think one of the reasons":** Michael Dinkins, "What Jack Welch Taught This CFO about Leadership," Spend Culture Stories Podcast, 2018, soundcloud.com/spendculture/what-it-was -like-to-work-with-jack-welch-michael-dinkins.

69 **"The opportunity to lead Walmart":** "Doug McMillon Elected New Chief Executive Officer of Wal-Mart Stores, Inc.," Walmart, November 25, 2013, corporate.walmart.com/_news_/news-archive /2013/11/25/doug-mcmillon-elected-new-chief-executive-officer-of -wal-mart-stores-inc.

CHAPTER 5: THE RESPONSIBILITY OF BUSINESS (REVISED)

71 **"In a free-enterprise":** Milton Friedman, "A Friedman Doctrine—The Social Responsibility of Business Is to Increase Its

Profits." *The New York Times Magazine*, September 13, 1970, www
.nytimes.com/1970/09/13/archives/a-friedman-doctrine-the-social
-responsibility-of-business-is-to.html.

71 **"only one social responsibility":** Friedman, "A Friedman
Doctrine."

72 **"Consumption . . . is the sole end":** Adam Smith, *The Wealth of
Nations*, Part Two (New York: Collier, 1902), 442.

75 **As Henry Ford said:** Quoted in Bryce G. Hoffman, *American
Icon: Alan Mulally and the Fight to Save Ford Motor Company*
(New York: Crown Publishing, 2012), 398.

79 **Thanks in large part:** Caroline Fohlin, "A Brief History of
Investment Banking from Medieval Times to the Present," in *The
Oxford Handbook of Banking and Financial History*, ed. Youssef
Cassis et al. (Oxford: Oxford University Press, 2014).

82 **They act more like renters:** Reference from Tom Staggs.

86 **Leaderless and unfocused:** Michael Levitin, "The Triumph of
Occupy Wall Street," *The Atlantic*, June 10, 2015. www
.theatlantic.com/politics/archive/2015/06/the-triumph-of-occupy
-wall-street/395408/; and Ray Sanchez, "Occupy Wall Street: 5
Years Later," CNN.com, September 16, 2016, www.cnn.com/2016
/09/16/us/occupy-wall-street-protest-movements/index.html.

89 **A People's War is:** Transcript of interview with Vo Nguyen Giap,
Viet Minh commander, *People's Century*, "Guerrilla Wars (1956–
1989)," season 1, episode 24, produced by BBC and WGBH
Boston, 1973, www.pbs.org/wgbh/peoplescentury/episodes
/guerrillawars/giaptranscript.html.

CHAPTER 6: WILL AND RESOURCES

96 **is 49 percent technical:** Diane Cardwell, "Spreading His Gospel
of Warm and Fuzzy," *The New York Times*, April 23, 2010, www
.nytimes.com/2010/04/25/nyregion/25meyer-ready.html.

98 **However, Costco, which pays their cashiers:** Andrés Cardenal,
"Higher Wages Could Pay Off for Wal-Mart Employees,
Customers, and Investors," *Motley Fool*, January 20, 2016, www
.fool.com/investing/general/2016/01/20/higher-wages-could-pay
-off-for-wal-mart-employees.aspx.

98 **because of that alternative mindset:** Zac Hall, "Retail Chief
Angela Ahrendts Talks 'Today at Apple' and More in Video
Interview," 9to5Mac, May 17, 2017, 9to5mac.com/2017/05/17
/angela-ahrendts-today-at-apple-video.

98 **average retention rates around 90 percent:** Don Reisinger,
"Here's How Apple's Retail Chief Keeps Employees Happy,"
Fortune, January 28, 2016, fortune.com/2016/01/28/apple
-retail-ahrendts-employees.

101   **money can't buy true will:** "Countless studies have shown that
       we're more committed to an activity when we do it out of passion,
       rather than an external reward such as a trophy." Jonathan Fader,
       PhD, "Should We Give Our Kids Participation Trophies?,"
       Psychology Today, November 7, 2014, www.psychologytoday.com
       /us/blog/the-new-you/201806/should-we-give-our-kids
       -participation-trophies.

101   **Tindell remembers what happened:** Author's interview with Kip
       Tindell.

CHAPTER 7: TRUSTING TEAMS

105   **Day after day:** Angus Chen, "Invisibilia: How Learning to Be
       Vulnerable Can Make Life Safer," NPR, June 17, 2016, www.npr
       .org/sections/health-shots/2016/06/17/482203447/invisibilia-how
       -learning-to-be-vulnerable-can-make-life-safer.

107   **"Part of safety . . . is being able":** Chen, "Invisibilia."

107   **And the results were remarkable:** Robin J. Ely and Debra
       Meyerson, "Unmasking Manly Men," *Harvard Business Review*,
       July–August 2008, hbr.org/2008/07/unmasking-manly-men.

107   **"I understand what you're saying":** Q & A session of the IACP
       conference in San Diego, 2016.

110   **Like the SEALs, Welch also ranked:** Personal interview with
       Navy SEAL Welch.

121   **"You *have* a problem":** Bryce G. Hoffman, *American Icon: Alan
       Mulally and the Fight to Save Ford* (New York: Crown Publishing,
       2012), 110–125.

121   **Once the Circle of Safety:** Author's interview with Alan Mulally

121   **"work together as a team":** Hoffman, *American Icon*, 121.

121   **Culture = Values + Behavior:** This formula was developed by
       Lt. Gen. George Flynn, USMC (ret).

CHAPTER 8: ETHICAL FADING

131   **As *The New York Times* reported:** Michael Corkery, "Wells
       Fargo Fined $185 Million for Fraudulently Opening Accounts,"
       The New York Times, September 8, 2016, www.nytimes.com/2016
       /09/09/business/dealbook/wells-fargo-fined-for-years-of-harm-to
       -customers.html.

132   **Ultimately, 5,300 Wells Fargo employees:** Chris Arnold,
       "Former Wells Fargo Employees Describe Toxic Sales Culture,
       Even at HQ," NPR, October 4, 2016, www.npr.org/2016/10/04
       /496508361/former-wells-fargo-employees-describe-toxic-sales
       -culture-even-at-hq.

135   **Some employees recall being pushed:** Arnold, "Former Wells
       Fargo Employees Describe Toxic Sales Culture."

136 **As another Wells Fargo employee confessed:** "Wells Fargo Workers Created Fake Accounts," video, CNN Business, April 10, 2017, money.cnn.com/2017/04/10/investing/wells-fargo-board -investigation-fake-accounts/index.html.

136 **Investigations into the scandal:** Matt Egan, "Wells Fargo Claws Back $75 Million from Former CEO and Top Exec," CNN Business, April 10, 2017, money.cnn.com/2017/04/10/investing /wells-fargo-board-investigation-fake-accounts/index.html; and Independent Directors of the Board of Wells Fargo & Company, "Sales Practices Investigation Report," April 10, 2017, www .documentcloud.org/documents/3549238-Wells-Fargo-Sales -Practice-Investigation-Board.html.

136 **by 2010, a year before:** Matt Egan, "Feds Knew of 700 Wells Fargo Whistleblower Cases in 2010," CNN Business, April 19, 2017, money.cnn.com/2017/04/19/investing/wells-fargo-regulator -whistleblower-2010-occ/index.html?iid=EL.

136 **John Stumpf became aware:** Independent Directors of the Board of Wells Fargo & Company, "Sales Practices Investigation Report," April 10, 2017, 55, www.documentcloud.org/documents/3549238 -Wells-Fargo-Sales-Practice-Investigation-Board.html.

136 **She was also, according to the report:** Independent Directors of the Board of Wells Fargo & Company, "Sales Practices Investigation Report," 13, 8, 46, www.documentcloud.org /documents/3549238-Wells-Fargo-Sales-Practice-Investigation -Board.html.

137 **In 2018, the bank was fined:** Julia Horowitz, "Wells Fargo to Pay $2.09 Billion Fine in Mortgage Settlement," CNN Business, August 1, 2018, money.cnn.com/2018/08/01/investing/wells-fargo -settlement-mortgage-loans/index.html.

137 **The auto division of the bank:** Emily Glazer, "Wells Fargo to Refund $80 Million to Auto-Loan Customers for Improper Insurance Practices," *The Wall Street Journal*, July 8, 2017, www .wsj.com/articles/wells-fargo-to-refund-80-million-to-auto-loan -customers-for-improper-insurance-practices-1501252927.

137 **And the wholesale division:** "U.S. Probing Wells Fargo's Wholesale Banking Unit: WSJ," Reuters, September 6, 2018, www.reuters.com/article/us-wells-fargo-probe/u-s-probing-wells -fargos-wholesale-banking-unit-wsj-idUSKCN1LM28O.

137 **To put things in perspective:** Wells Fargo & Company Annual Report 2016, 37, 88.3B.

137 **Indeed, John Stumpf did lose his job:** Matt Krantz, "Wells Fargo CEO Stumpf Retires with $134M," *USA Today*, October 13, 2016, www.usatoday.com/story/money/markets/2016/10/12/wells -fargo-ceo-retires-under-fire/91964778/.

138   **Seeing the impact these price increases had:** Mark Maremont,
"EpiPen Maker Mylan Tied Executive Pay to Aggressive Profit
Targets," *The Wall Street Journal*, September 1, 2016, www.wsj.com
/articles/epipen-maker-mylan-tied-executive-pay-to-aggressive
-profit-targets-1472722204; Aimee Picchi, "Mylan Boosted EpiPen's
Price Amid Bonus Target for Execs," CBS News, September 1,
2016, www.cbsnews.com/news/mylan-boosted-epipens-price-amid
-bonus-target-for-execs; and Gretchen Morgenson, "EpiPen Price
Rises Could Mean More Riches for Mylan Executives," *The New
York Times*, September 1, 2016, www.nytimes.com/2016/09/04
/business/at-mylan-lets-pretend-is-more-than-a-game.html.

139   **No doubt responding to this incentive:** Morgenson, "EpiPen
Price Rises Could Mean More Riches for Mylan Executives."

139   **After the fifteenth price hike since 2009:** Catherine Ho, "CEO
at Center of EpiPen Price Hike Controversy Is Sen. Joe Manchin's
Daughter," *The Washington Post*, August 24, 2016, www
.washingtonpost.com/news/powerpost/wp/2016/08/24/ceo-at
-center-of-epipen-price-hike-controversy-is-sen-joe-manchins
-daughter/?utm_term=.7f474849840b; and Matt Egan, "How
EpiPen Came to Symbolize Corporate Greed," CNN Business,
August 29, 2016, money.cnn.com/2016/08/29/investing/epipen
-price-rise-history/index.html.

139   **CEO Heather Bresch replied:** Danielle Wiener-Bronner, "Mylan
CEO: You Can't Build a Company in a Quarter," CNN Business,
June 4, 2018, money.cnn.com/2018/06/04/news/companies
/heather-bresch-boss-files/index.html.

140   **William D. Weinreb explained:** "Mylan Agrees to Pay $465
Million to Resolve False Claims Act Liability for Underpaying
EpiPen Rebates," U.S. Department of Justice, Office of Public
Affairs, August 17, 2017, www.justice.gov/opa/pr/mylan-agrees
-pay-465-million-resolve-false-claims-act-liability-underpaying
-epipen-rebates.

140   **David Messick, professor emeritus:** Ann E. Tenbrunsel and
David M. Messick, "Ethical Fading: The Role of Self-Deception
in Unethical Behavior," Social Justice Research 17, no. 2 (June
2004): 223–36.

143   **Tenbrunsel and Messick identify the proverbial "slippery
slope":** Tenbrunsel and Messick, "Ethical Fading," 228–29.

145   **Tim Sloan admitted that:** Matt Egan, "Elizabeth Warren to
Wells Fargo CEO: 'You Should Be Fired,'" CNN Business,
October 3, 2017, money.cnn.com/2017/10/03/investing/wells-fargo
-hearing-ceo; and "Wells Fargo Statement Regarding Board
Investigation into the Community Bank's Retail Sales Practices,"
Business Wire, April 10, 2017, www.businesswire.com/news/home

/20170410005754/en/Wells-Fargo-Statement-Board-Investigation
-Community-Bank%E2%80%99s.

147 **"process will always tell us":** "Dr. Leonard Wong Discusses a
Culture of Dishonesty in the Army," STEM-Talk, Episode 29,
Florida Institute for Human & Machine Cognition, January 17,
2017, www.ihmc.us/stemtalk/episode-29-2.

147 **However, in their paper:** Leonard Wong and Stephen J. Gerras,
"Lying to Ourselves: Dishonesty in the Army Profession," U.S.
Army War College Strategic Studies Institute (Carlisle Barracks,
PA: U.S. Army War College Press, 2015), ssi.armywarcollege.edu
/pdffiles/pub1250.pdf.

148 **One example Wong and Gerras give:** Wong and Gerras, "Lying
to Ourselves."

152 **The copy read:** Tim Nudd, "Ad of the Day: Patagonia," *Adweek*,
November 28, 2011, www.adweek.com/brand-marketing/ad-day
-patagonia-136745/.

152 **"We did it out of guilt":** Monte Burke, "The Greenest
Companies in Fly Fishing," FlyFisherman.com, February 1, 2016,
www.flyfisherman.com/conservation/greenest-companies-in-fly
-fishing/Chouinard.

153 **"We plan to be here":** Katya Margolin, "Could Patagonia's
Alternative Leadership Model Unleash the Best in Your People?,"
Virgin, October 7, 2016, www.virgin.com/entrepreneur/could
-patagonias-alternative-leadership-model-unleash-best-your-people.

153 **Patagonia is driven by a vision:** Patagonia, "Sustainability
Mission/Vision," https://www.patagonia.com/sustainability.html.

154 **As their website says:** Patagonia, "Don't Buy This Jacket, Black
Friday and the *New York Times*," November 25, 2011, www
.patagonia.com/blog/2011/11/dont-buy-this-jacket-black-friday-and
-the-new-york-times.

155 **As a result of separate internal audits:** Gillian B. White, "All
Your Clothes Are Made with Exploited Labor," *The Atlantic*, June
3, 2015, www.theatlantic.com/business/archive/2015/06/patagonia
-labor-clothing-factory-exploitation/394658.

156 **"The pressure of a public company":** Margolin, "Could
Patagonia's Alternative Leadership Model Unleash the Best in
Your People?"

156 **To certify as a B Corp:** "Three Guides for Going B—and Why It
Matters," Patagonia.com, August 27, 2018, www.patagonia.com
/blog/2018/08/three-guides-for-going-b-and-why-it-matters.

157 **In the words of Patagonia's CEO:** Jeff Beer, "How Patagonia
Grows Every Time It Amplifies Its Social Mission," *Fast
Company*, February 21, 2018, www.fastcompany.com/40525452
/how-patagonia-grows-every-time-it-amplifies-its-social-mission.

157 **"If we can show the business community"**: "Clothing Company Tells Customers to Buy Less," *PBS NewsHour*, August 21, 2015, www.pbs.org/newshour/extra/daily-videos/clothing-company-tells -consumers-to-buy-less.

CHAPTER 9: WORTHY RIVAL

162 **Chris Evert Lloyd and Martina Navratilova:** Gwen Knapp, "Evert vs. Navratilova—What a Rivalry Should Be," *San Francisco Chronicle*, June 19, 2005, www.sfgate.com/sports/knapp/article /Evert-vs-Navratilova-what-a-rivalry-should-be-2661371.php.

162 **It is the focus on process and constant improvement:** "Countless studies have shown that we're more committed to an activity when we do it out of passion, rather than an external reward such as a trophy." Jonathan Fader, PhD, "Should We Give Our Kids Participation Trophies?," *Psychology Today*, November 7, 2014, www.psychologytoday.com/us/blog/the-new-you/201806 /should-we-give-our-kids-participation-trophies.

164 **In the 15 years before Mulally took over:** Bryce G. Hoffman, *American Icon: Alan Mulally and the Fight to Save Ford* (New York: Crown Publishing, 2012), 109; and Sarah Miller Caldicott, "Why Ford's Alan Mulally Is an Innovation CEO for the Record Books," *Forbes*, June 25, 2014, www.forbes.com/sites/sarahcaldicott/2014 /06/25/why-fords-alan-mulally-is-an-innovation-ceo-for-the -record-books/#6b2caf297c04.

164 **One of the things he learned:** Hoffman, *American Icon*, 127.

164 **Detroit's car companies, including Ford:** Hoffman, *American Icon*, 97–98.

165 **"We're not going to chase market share":** Hoffman, *American Icon*, 139.

168 **Apple ran a full-page ad:** Bill Murphy Jr., "37 Years Ago, Steve Jobs Ran Apple's Most Amazing Ad. Here's the Story (It's Almost Been Forgotten)," Inc.com, August 23, 2018, www.inc.com/bill -murphy-jr/37-years-ago-steve-jobs-ran-apples-most-amazing-ad -heres-story-its-almost-been-forgotten.html.

168 **"Welcome to the most exciting":** Murphy Jr., "37 Years Ago, Steve Jobs Ran Apple's Most Amazing Ad."

174 **"The more I questioned these guys":** John Douglas, *Mindhunter: Inside the FBI's Elite Serial Crime Unit* (New York: Pocket Books, 1996), 56.

CHAPTER 10: EXISTENTIAL FLEXIBILITY

182 **It was, as his older brother Roy:** Neal Gabler, *Walt Disney: The Triumph of the American Imagination* (New York: Vintage, 2007), 10.

183 **the culture of Walt Disney Productions:** Gabler, *Walt Disney*, 492.

194 **The leaders of a once great infinite company:** Tendayi Viki, "On the Fifth Anniversary of Kodak's Bankruptcy, How Can Large Companies Sustain Innovation?," Forbes, January 19, 2017, www .forbes.com/sites/tendayiviki/2017/01/19/on-the-fifth-anniversary -of-kodaks-bankruptcy-how-can-large-companies-sustain -innovation/#5eb918e46280.

CHAPTER 11: THE COURAGE TO LEAD

196 **"But don't you sell cigarettes":** Larry Merlo, "The Good and the Growth in Quitting," TED Talk, Wake Forest University, YouTube video, 15:24, April 2015, www.youtube.com/watch?v= aM2ZtpqwYQs.

197 **"It might make money in Oz":** Jeff Morganteen, "Cramer: CVS' Tobacco Move Won't Fly on Wall Street," CNBC, February 5, 2015, www.cnbc.com/2014/02/05/cramer-cvs-tobacco-move-wont -fly-on-wall-street.html.

197 **One Illinois-based sales and marketing consultant:** "CVS' Tobacco Exit Draws Reaction, Applause," *Convenience Store News*, February 6, 2014, csnews.com/cvs-tobacco-exit-draws-reaction -applause.

197 **An independent study commissioned by CVS:** "We Quit Tobacco, Here's What Happened Next," Thought Leadership, CVS Health Research Institute press release, September 1, 2015, cvshealth.com/thought-leadership/cvs-health-research-institute /we-quit-tobacco-heres-what-happened-next.

198 **Companies like Irwin Naturals:** Brian Berk, "CVS Pharmacy Unveils the 'Next Evolution of the Customer Experience," Drug Store News, April 19, 2017, www.drugstorenews.com/beauty/cvs -pharmacy-unveils-next-evolution-customer-experience.

198 **CVS's stock price did fall:** Andrew Meola, "Rite Aid (RAD) and Walgreen (WAG) Rise on CVS Caremark (CVS) Tobacco Announcement," TheStreet, February 5, 2014, www.thestreet.com /story/12311827/1/rite-aid-rad-and-walgreen-wag-rise-on-cvs -caremark-cvs-tobacco-announcement.html.

199 **Prior to the announcement:** CVS Health Corporation Revenue & Earnings Per Share (EPS), Nasdaq, data as of April 2, 2019, www.nasdaq.com/symbol/cvs/revenue-eps.

201 **"She told me I couldn't let our airline fail":** Author's interview with Doug Parker.

205 **The scandal ultimately resulted in:** Edward Wong, "Under Fire for Perks, Chief Quits American Airlines," *The New York Times*, April 25, 2003, www.nytimes.com/2003/04/25/business/under-fire -for-perks-chief-quits-american-airlines.html.

205 **"Saying you trust people is just words":** Author's interview with Doug Parker.

206 **And seeing their employees left behind:** "A Letter to American Employees from Doug Parker and Robert Isom on Team Member Pay," American Airlines Newsroom, April 26, 2017, news.aa.com /news/news-details/2017/A-letter-to-American-employees-from -Doug-Parker-and-Robert-Isom-on-team-member-pay/default.aspx.

206 **One analyst, Kevin Crissey:** "The Case Against 'Maximizing Shareholder Value,'" NPR, May 6, 2017, www.npr.org/2017/05/06 /527139988/the-case-against-maximizing-shareholder-value.

206 **"We are troubled by AAL's wealth transfer":** John Biers, "American Airlines Defends Pay Increase As Shares Tumble," Yahoo Finance, April 27, 2017, finance.yahoo.com/news/american -airlines-boosts-employee-pay-earnings-fall-130526168.html.

207 **We "have to work on it":** Author conversation with Doug Parker, May 3, 2019.

208 **"The worst sort of business":** Quoted in Adam Levine-Weinberg, "7 Ways Warren Buffett Blasted the Airline Industry—Before Investing Billions There," *Motley Fool*, March 5, 2017, www.fool .com/investing/2017/03/05/7-ways-warren-buffett-blasted-the -airline-industry.aspx.

209 **"help people across the world":** Walgreens Boots Alliance, "About us," "Vision, Purpose and Values," www.walgreens bootsalliance.com/about/vision-purpose-values.

209 **"active decision to reduce space":** Ronnie Cohen, "When CVS Stopped Selling Cigarettes, Some Customers Quit Smoking," Reuters, Health News, March 20, 2017, www.reuters.com/article /us-health-pharmacies-cigarettes/when-cvs-stopped-selling -cigarettes-some-customers-quit-smoking-idUSKBN16R2HY.

209 **"We've reviewed this on a regular basis":** Lisa Schencker, "Why Is Walgreens Still Selling Cigarettes? Shareholders Want to Know," *Chicago Tribune*, January 26, 2017, www.chicagotribune .com/business/ct-walgreens-selling-cigarettes-0127-biz-20170126 -story.html.

210 **According to the Centers for Disease Control:** "Health Effects of Cigarette Smoking," Centers for Disease Control and Prevention, www.cdc.gov/tobacco/data_statistics/fact_sheets /health_effects/effects_cig_smoking/index.htm.

210 **All those smoking-related illnesses:** "Economic Trends in Tobacco," Centers for Disease Control and Prevention, www.cdc .gov/tobacco/data_statistics/fact_sheets/economics/econ_facts /index.htm.

210 **Nearly 70 percent of all smokers:** Centers for Disease Control Prevention (CDC), "Quitting Smoking Among Adults—United States, 2001–2010," MMWR. Morbidity and Mortality Weekly

Report 60, no. 44 (November 11, 2011): 1513–19, www.ncbi.nlm
.nih.gov/pubmed?term=22071589; https://www.cdc.gov/tobacco
/data_statistics/fact_sheets/cessation/quitting/index.htm

212 **"At Rite Aid, we have":** Rite Aid, "About Us," "Our Story," www
.riteaid.com/about-us/our-story.

212 **"Rite Aid offers a wide range":** Paul Edward Parker, "Rite Aid
Responds to CVS Decision to Stop Selling Tobacco," *Providence
Journal*, February 6, 2014, www.providencejournal.com/breaking
-news/content/20140206-rite-aid-responds-to-cvs-decision-to-stop
-selling-tobacco.ece.

213 **The day after CVS made the announcement:** CVS Health, CVS
Purpose Short, YouTube, October 9, 2017, www.youtube.com
/watch?v=Geq6HuItPN4.

214 **"They wait for public opinion":** Steve Lohr and Landon Thomas
Jr., "The Case Some Executives Made for Sticking with Trump,"
The New York Times, August 17, 2017, www.nytimes.com/2017/08
/17/business/dealbook/as-executives-retreated-lone-voices-offered
-support-for-trump.html.

INDEX

SIMON SINEK

START WITH WHY

Based on the most-watched TED Talk of all time.

Why are some people and organizations more inventive, pioneering and successful than others? And why are they able to repeat their success again and again?

Because in business it doesn't matter what you do, it matters <u>why</u> you do it.

Steve Jobs, the Wright brothers and Martin Luther King have one thing in common: they STARTED WITH WHY.

This book is for anyone who wants to inspire others, or to be inspired.

'One of the most useful and powerful books I have read in years'
William Ury, co-author of *Getting to Yes*

'This book is so impactful, I consider it required reading' Tony Robbins,
bestselling author of *Awaken the Giant Within*

PENGUIN PARTNERSHIPS

Penguin Partnerships is the Creative Sales and Promotions team at Penguin Random House. We have a long history of working with clients on a wide variety of briefs, specializing in brand promotions, bespoke publishing and retail exclusives, plus corporate, entertainment and media partnerships.

We can respond quickly to briefs and specialize in repurposing books and content for sales promotions, for use as incentives and retail exclusives as well as creating content for new books in collaboration with our partners as part of branded book relationships.

Equally if you'd simply like to buy a bulk quantity of one of our existing books at a special discount, we can help with that too. Our books can make excellent corporate or employee gifts.

Special editions, including personalized covers, excerpts of existing books or books with corporate logos can be created in large quantities for special needs.

We can work within your budget to deliver whatever you want, however you want it.

**For more information, please contact
salesenquiries@penguinrandomhouse.co.uk**